T0225993

Lecture Notes in Computer Science   **9089**

Commenced Publication in 1973
Founding and Former Series Editors:
Gerhard Goos, Juris Hartmanis, and Jan van Leeuwen

More information about this series at http://www.springer.com/series/7409

Thanassis Tiropanis · Athena Vakali
Laura Sartori · Pete Burnap (Eds.)

Internet Science

Second International Conference, INSCI 2015
Brussels, Belgium, May 27–29, 2015
Proceedings

 Springer

Editors

Thanassis Tiropanis
University of Southampton
Southampton
UK

Athena Vakali
Aristotle University of Thessaloniki
Thessaloniki
Greece

Laura Sartori
University of Bologna
Bologna
Italy

Pete Burnap
Cardiff University
Cardiff
UK

ISSN 0302-9743   ISSN 1611-3349   (electronic)
Lecture Notes in Computer Science
ISBN 978-3-319-18608-5   ISBN 978-3-319-18609-2   (eBook)
DOI 10.1007/978-3-319-18609-2

Library of Congress Control Number: 2015937738

LNCS Sublibrary: SL3 – Information Systems and Applications, incl. Internet/Web and HCI

Springer Cham Heidelberg New York Dordrecht London

Printed on acid-free paper

Springer International Publishing AG Switzerland is part of Springer Science+Business Media
(www.springer.com)

Preface

The 2nd International Conference on Internet Science "Societies, governance and innovation" took place in Brussels during May 27–29, 2015, under the aegis of the European Commission, by the EINS project, the FP7 European Network of Excellence in Internet Science.

The main objective of this highly multidisciplinary conference is to foster dialogue among scholars and practitioners in Internet science; an interdisciplinary area that draws on Computer Science, Sociology, Art, Mathematics, Physics, Complex systems analysis, Psychology, Economics, Law, Political Science, Epistemology, and other disciplines in order to study the Internet as a sociotechnical phenomenon closely related to various areas of human activity, under technological and humanistic perspectives. This year the focus was on the contribution and role of Internet science on the current and future multidisciplinary understanding of societal transformation, governance shifts, and innovation quests.

This conference was built on the achievements of the 1st International Conference on Internet Science, which was successfully held during April 9–11, 2013 at the Royal Flemish Academy of Belgium for Science and the Arts, Brussels, organized by the EINS project, with the support of the European Commission DG CONNECT and attracted over 140 participants.

The accepted papers of the 2nd International Conference on Internet Science are organized into the following thematic topics:

- Internet Science in Reflection
- Internet Science and Societal Innovations
- Internet and Innovation

May 2015

Thanassis Tiropanis
Athena Vakali
Laura Sartori
Pete Burnap

Organization

Executive Committee

General Chair

Thanassis Tiropanis   University of Southampton, UK

Program Committee Co-chairs

Athena Vakali	Aristotle University of Thessaloniki, Greece
Laura Sartori	University of Bologna, Italy

Organizing Co-chairs

Anna Satsiou	Centre for Research and Technology Hellas, Greece
Franco Bagnoli	University of Florence, Italy
Hugo Vivier	Sigma Orionis, France
Cindy Temps	Sigma Orionis, France

Event Coordinator

Roger Torrenti   Sigma Orionis, France

Proceedings Chair

Pete Burnap   Cardiff University, UK

Program Committee

Andreas Fischer	University of Passau, Germany
Anna Satsiou	Centre for Research and Technology Hellas, Greece
Anne-Marie Oostveen	University of Oxford, UK
Armin Haller	Australian National University, Australia
Barbara Catania	University of Genoa, Italy
Bart Lannoo	iMinds, Belgium
Ben Zevenbergen	University of Oxford, UK
Clare Hooper	University of Southampton, UK

Chris Marsden	University of Sussex, UK
Dimitri Papadimitriou	Alcatel Lucent, Bell Labs, Belgium
Donny McMillan	Mobile Life, Sweden
Federico Morando	Nexa Center for Internet and Society, Italy
George Iosifidis	Centre for Research and Technology Hellas, Greece
Heiko Niedermeyer	Technical University of Munich, Germany
Ioannis Stavrakakis	University of Athens, Greece
Jonathan Cave	University of Warwick, UK
Karmen Guevarra	Cambridge University, UK
Konstantinos Kafetsios	University of Crete, Greece
Mahdi Bohlouli	University of Siegen, Germany
Mark Rouncefield	University of Lancaster, UK
Merkourios Karaliopoulos	Centre for Research and Technology Hellas, Greece
Mayo Fuster Morell	Harvard University, USA
Meryem Marzouki	University Pierre and Marie Curie, France
Nathalie Mitev	London School of Economics and Political Science, UK
Panayotis Antoniadis	ETH Zürich, Switzerland
Patrick Wüchner	University of Passau, Germany
Pete Burnap	Cardiff University, UK
Sara Helen Wilford	De Montfort University, UK
Tamas David-Barrett	University of Oxford, UK
Tim Chown	University of Southampton, UK
Tim Davies	University of Oxford, UK
Xin Wang	University of Southampton, UK
Ziga Turk	University of Ljubljana, Slovenia

Contents

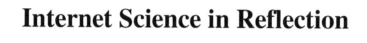

Internet Science in Reflection

Privacy and Empowerment in Connective Media

Jo Pierson[(✉)]

Vrije Universiteit Brussel (iMinds-SMIT), Brussels, Belgium
jo.pierson@vub.ac.be

Abstract. In order to explore the roadmap and future directions of social, mobile and ubiquitous media technologies for privacy from an Internet Science perspective we take an interdisciplinary perspective, building on research in Media and Communication Studies and Science and Technology Studies. In order to assess to what extent and how people are effectively 'empowered' in relation to their privacy, we investigate how they socially engage with these media and how their communication is being (re)configured. For this we first take a macro perspective by discussing recent interdisciplinary views on the changing internet landscape. Next we take a micro perspective on the position of users of 'connective media'. We then explain and situate user empowerment in relation to online privacy and how this is related to vulnerability. This perspective is used for discussing (young) people and their privacy management, based on the findings of online privacy research projects.

Keywords: Privacy · Empowerment · Connective media · Interdisciplinarity · Vulnerability

1 Socio-Technical Context

1.1 Culture of Connectivity

Institutional economists Freeman and Soete [1] and Perez [2] identify the current era – since the end of 20th century – as the fifth techno-economic paradigm shift, based on computing and information and communication technology (ICT). The current wave of technological invention and economic disruption, with its hallmarks of computer power, connectivity and data ubiquity, promises to deliver a similar mixture of social stress and economic transformations like former industrial revolutions [3: 3]. More in particular two big socio-technological transformations have taken place in the past decade, which have substantially affected our experience of sociality: (1) the transformation from a participatory culture to a culture of connectivity and (2) the transformation from networked communication to a 'platformed' sociality [4].

The concept of 'participatory culture' was born in the 1990s. It reflects the Internet's potential to nurture connections, build communities and advance democracy. It was a need for connectedness that drove users to the web in the first place. In the beginning of the new millennium more and more people started using websites for making and maintaining connections, by sharing creative content and

© Springer International Publishing Switzerland 2015
T. Tiropanis et al. (Eds.): INSCI 2015, LNCS 9089, pp. 3–14, 2015.
DOI: 10.1007/978-3-319-18609-2_1

enjoying their social lives online. As people's lives became permeated with social media platforms, they started to move their social, cultural and professional activities to an online environment. Van Dijck [4: 4-5] notices that existing or new media and technology companies incorporated the existing social platforms. These companies were not so much interested in the ideals of the participatory culture, as they were in the (personal) data that the users delivered, as a by-product of maintaining connections online. They made use of algorithms that engineer and manipulate the social connections. The latter is labeled as 'connectivity': an automated process behind the real-life connections, which made it possible to recognize people's desires. As such, a profitable form of sociality has been created. Therefore it would be better to use the term 'connective media' instead of 'social media'. As stated by van Dijck [4: 13-14] "What is claimed to be 'social' is in fact the result of human input shaped by computed output and vice versa – a sociotechnical ensemble whose components can hardly be told apart. The norms and values supporting the 'social' image of these media remain hidden in platforms' technological textures". The first transformation goes hand in hand with the second one: from networked communication to 'platformed' sociality. "It is a common fallacy, though, to think of platforms as merely facilitating networking activities; instead, the construction of platforms and social practices is mutually constitutive." [4: 6]. As the online platforms are no longer sheer carriers for communication, human sociality is being brought to these platforms and at the same time mediated by them.

So the culture of connectivity is a culture where perspectives, expressions, experiences and productions are increasingly mediated by digital media and their algorithms [5]. It is this mediation and manipulation of social relationships and the gathering of people's preferences that influenced the privacy of individuals online. Mere outings of sociality online have become structured and tracked; they are released on an electronic platform, which can have far-reaching and long-lasting effects [5: 7].

1.2   Mass Self-Communication

The transitions in this culture of connectivity on macro level go together with a shift from mass media and personal media to media for 'mass self-communication' on micro level [6]. Castells [7: 55] sees the latter as the new kind of communication in contemporary society. On the one hand mass communication because social media can potentially reach a worldwide internet audience. On the other hand 'self-communication' because the message production is typically self-generated, the potential receiver(s) definition is self-directed and the message or content retrieval is self-selected. The different forms of communication (mass media, interpersonal communication and mass self-communication) complement rather than substitute each other.

The rise of mass self-communication also intensifies the move towards 'networked individualism' where the individual person becomes the portal or hub for different networks around him or her [8]. This gives media producers/distributors/consumers a lot of freedom and leverage in how, when, what and with whom communication takes place. However at the same time the increased level of individual control often implies more responsibilities given the possible serious consequences.

1.3   User (Dis)Empowerment

These pros and cons are linked to notions of respectively 'user empowerment' and 'user disempowerment'. Empowerment is a widely used concept charged with meaning. It has a long tradition in social sciences, more in particular in social welfare and civil society literature, but also in science, business and policy fields. User empowerment is described by Zimmerman and Rappaport [9] as the process of strengthening individuals, by which they get a grip on their situation and environment, through the acquisition of more control, sharpening their critical awareness and the stimulation of participation. In the saturated media environment user empowerment can be defined as the capability for interpreting and acting upon the social world that is intensively mediated by social media [10: 409]. User empowerment is then dependent on knowledge of how mechanisms operate and from what premise, as well as on the skills to change them [4: 171].

In a culture of connectivity where our social lives are increasingly mediated by mass self-communication, people do not always own the necessary capabilities to optimally interpret and act upon other people and institutions for acquiring an equal position in society [6: 104]. In this lies a risk of disempowerment that is visible in issues of digital media and privacy.

1.4   Vulnerability

As mentioned before the unprecedented autonomy of media users and hence increased self-directed control over time, place and content of communication and interaction with many more people, increases the chance of positive – but also negative – consequences and implies more responsibilities. There is for example a substantial chance that online user practices via social media are more persistent in time, have a broader geographical reach and are picked up by unwanted receivers. This means that the 'vulnerability' of people engaging in mass self-communication changes and possibly increases, which intensifies the need for empowerment and aggravates the risk of disempowerment. Where vulnerability defines the circumstances of potential risk as they are, disempowerment refers to people actually losing power and capabilities to gain control over their lives mediated by social media.

The concept of 'vulnerability' (and its opposite 'security') has been intensively discussed in the studies on human development, geography, disaster reduction, and risk communication [11]. It is often – wrongfully – equaled with 'poverty', but it in fact it has (or can have) a much broader meaning. However, vulnerability has not yet a developed theory and accepted indicators and methods of measurement, though Chambers [12] sees an external side of vulnerability related to 'exposure' (see also Ball [13: 647]) and an internal side related to 'coping capacities'. Watts and Bohle [14] and Bohle [15] have further expanded this differentiation, keeping the structure of external and internal sides of vulnerability. They have defined vulnerability as a multi-layered and multi-dimensional social space defined by the political, economic, and institutional capabilities of people in specific places and times. The external perspective refers mainly to the structural dimensions of vulnerability exposure, while the internal dimension of vulnerability focuses on coping and action to overcome or at least mitigate negative effects [15].

This twofold approach of vulnerability is also reflected in the way social media technologies have been approached from a Science and Technology Studies (STS) perspective, confronting the structural element of 'affordances' with the action-oriented element of 'practices'. Exposure to vulnerability results from the 'affordances', defined as the combination of perceived and actual properties of the (social media) technology, primarily those fundamental properties that determine just how that technology could possibly be used [16, 17]. Coping with vulnerability happens in the 'practices', defined as 'recognisable entities', but at the same time "require constant and active reproduction or performance" [18]. In this way a 'practice' is seen as a routinised type of behaviour. From an STS perspective there is no essential use to be derived from the technological (social media) artefact itself, because technologies should be studied in their context of user practices and users and technologies should be seen as co-constructed [19]. We use the structure of external and internal sides of vulnerability and the related affordances and practices of social media, for our further analysis. For this we take a closer look at how people's vulnerability (and possible disempowerment) is reconfigured within the changing media landscape of mass self-communication. To illustrate these transitions, we focus on privacy issues in a culture of connectivity.

In the next paragraphs we discuss key empirical findings from three online privacy research projects.

- USEMP (User empowerment for Enhanced Online Management) in EU (http://www.usemp-project.eu), a three-year STREP research project funded by European Commission 7th Framework Programme for research, technological development and demonstration (Grant agreement no. 611596) (2013-2016).
- SoLoMidem (SOcial, LOcal & Mobile IDEntity Management) in Flanders/Belgium (http://www.iminds.be/en/projects/2014/03/18/solomidem), a two-year interdisciplinary ICON research project, supported by iMinds and the Flemish Agency for Innovation by Science and Technology (2012-2014).
- SPION (Security and Privacy for Online Social Networks) in Flanders/Belgium (www.spion.me), a four-year interdisciplinary Strategic Basic Research research project (SBO) by four universities, funded by IWT (2010-2014).

By integrating results from these projects we are able to demonstrate the changing vulnerability with regards to exposure to and coping with online privacy. The interdisciplinary theoretical background is based on combining insights from Media and Communication Studies (MCS), STS and sociology.

2   Vulnerability in Managing Online Privacy

2.1   Exposure to Online Privacy: Case Study on Location-Based Social Networks and Privacy[1]

We first look at the way that users are exposed to (risks of) online privacy in different internet and media technologies. For this we highlight privacy scripts in location-based social networks. With the increasing adoption of smartphones, location-based social networks and applications (e.g. Foursquare, Find My Friends) gain widespread popularity. However, the disclosure of location information within these networks can cause privacy concerns among mobile users. In most of the research on privacy in location-based social networks, technology is researched as a context factor for explaining privacy related behaviour. However in the SoLoMidem project we noticed – from an STS perspective – how the privacy situation is formatted in the relation between so-called technical scripts and user practices. Following the work of Akrich [20], we studied the privacy scripts in two location-based social networks. In a qualitative user study we researched their framework of action and how they shape privacy concerns.

Introduction
A technology can shape a framework of action, but users might also use the technology in a way not foreseen by the designer. Akrich [20] refers to this as the script or the scenario of a technology. The design of a technology defines what decisions can be made by the user, and what is controlled by the 'machine'. Technologies define a framework of action for the user, the space in which they are supposed to act and way in which they interact. We apply these insights to study the Location-Based Social Networks (LBSN) users' (location) privacy concerns. To explore how these concerns can guide privacy practices, we also have to scrutinize the LBSN and how the user is inscribed or scripted into these technologies. Users nowadays make privacy decisions in a not so transparent market. In their search for creating a good user experience and a valid business model, different location-based service providers embed the user location sharing practices into their applications in different ways. We study the scripts of LBSN and how users' privacy concerns exist in the framework of action of these scripts.

Research Set-up
The study was done in 2013 and had two parts: a comparative analysis of the privacy scripts of two commercial LBSN (Foursquare and Glympse) and a qualitative user study (field trial) in which we study how the privacy scripts of these LBSN shape privacy concerns.

---

[1] Based on: Coppens, Paulien, Claeys, Laurence, Veeckman, Carina & Pierson, Jo (2014) Privacy in location-based social networks: Researching the interrelatedness of scripts and usage, Paper at SOUPS 2014 - Symposium On Usable Privacy and Security - Workshop on Privacy Personas and segmentation (PPS), 9-11 July 2014, Menlo Park (CA), US.

Findings

The analysis of the privacy scripts of Foursquare (www.foursquare.com), a check-in service, and Glympse (www.glympse.com), a location tracking service, demonstrate that the two LBSN have very different privacy scripts. Glympse is an application designed to give users almost full control over their location sharing settings. Foursquare on the other hand, defines another framework of action for the user. It appears to be in the interest of the Foursquare application, contrary to Glympse, that as many personal data or settings are set as public. For a rather big set of data, users have no possibility to set it as private and the privacy settings that are adjustable, are set public by default. A possible explanation for this is the fact that Foursquare works together with local businesses and brands that have access to aggregated and anonymous data. We could thus say that, contrary to Glympse, the preferred reading of Foursquare limits users in the management of online privacy.

The user study focused on how the privacy scripts of Foursquare and Glympse shape participants' privacy concerns and practices and on users' coping mechanisms in redefining this framework of action. The comparative analysis of the privacy scripts revealed that Glympse is a more privacy preserving LBSN. However, the respondents of our study perceived Glympse to be more privacy-invasive than Foursquare. Glympse is a location-tracking service, typically causing higher privacy concerns than the position-aware service Foursquare (see also Barkhuus and Dey [21]). On top of that, we have to take into account the control factor. Although the analysis of the privacy scripts shows that Glympse gives users more control mechanisms (e.g. control over audience), most of the participants felt they have a higher control over the disclosure of their location when using Foursquare. Important here, contrary to actual control, is thus the notion of perceived control or the "illusion of control" (see also Brandimarte et al. [22]). Within the check-in service Foursquare, they each time can to make the conscious decision to share a specific location, while in Glympse a longer location path is shared. When taking a closer look at the participants' practices, we can see that, although the importance of control over location data is continuously stressed, they do not always act upon this. However, we should also note that when a LBSN does not offer enough control features to protect location information, the participants of our study sometimes applied very inventive strategies to protect their location to compensate the lack of control. An example here is a participant who reports to never check in with Foursquare every day and/or at the same time at work and at home to not make it easy for someone with malicious intents to discover daily routines.

Although respondents say they want full control over privacy settings, our study shows that they are not willing to each time define elaborate sharing settings. This also continuously makes the process of sharing personal information more visible for users, and users do not always like to be confronted with that. The challenge for LBSN providers is to find the right balance between privacy-preserving affordances and an optimal user experience.

The study shows that Glympse responds to mobile users' privacy concerns by giving them more options to control the disclosure of location information. Foursquare defines another framework of action for the user. It appears to be in the interest of Foursquare that users provide many personal data that are set as public. This is illustrated, among others, by the fact that, although all the respondents want high control over the (location) privacy settings, almost none of them ever checked Foursquare's privacy settings before. However when thoroughly investigating the usage, Foursquare is considered by users as less privacy invasive than Glympse. This is partly due to the properties of the LBSN. Glympse is a location-tracking service, which decreases the respondents' feeling of control. But also to the fact that the check-in service Foursquare makes it easier for the respondents to apply their own strategies to control location disclosure, hereby opposing the application's framework of action.

2.2   Coping with Online Privacy: Case Study on Adolescents, Young Adults and Intergroup Privacy Management[2]

In order to have a complete picture of the vulnerability in managing online privacy, we need to complement exposure by media technologies with coping capabilities of users. On the latter level we focus on young people, as they are the first generation that have incorporated digital media into their everyday life. In this way we are able to see how these young people are indeed more vulnerable – as is often claimed – or are able to cope with issues of online privacy.

For this we have investigated adolescents, in order to understand how they are dealing with online disclosure individually and in group. This is especially relevant as the properties of social media challenge young people's boundary coordination and privacy management. Most research on privacy management within the context of Online Social Networks (OSNs) treats the people as individual owners of private information. Privacy, however, is beyond individual control and is also managed on an intergroup level. In order to investigate this we have studied how youth organizations negotiate their offline and online privacy. The theoretical framework of Goffman [23] and the communication privacy management theory [24] are used to frame privacy management and boundary coordination. The results indicate how spatial, temporal, and social boundaries are lengthened and translated online to maintain privacy. The collapse of boundaries in OSN, however, forces youth organizations to adopt explicit privacy rules that are routinized and more implicitly present in an offline environment, and creates tension between individual and group privacy rules.

Introduction
Through interacting with others and the self, individuals create various types of identities. Adhering to the symbolic interactionist viewpoint in sociology, identity formation is not an individual project but a socially constructed and negotiated one, which constitutes an essential component of the self [25]. On the one hand,

---

[2] Based on: De Wolf, Ralf & Pierson, Jo (2014) Security and Privacy for Online Social Networks (SPION): Deliverable 7.7 – User practices profile on everyday technologically and non-technologically privacy practices & Deliverable 7.8 – Identification of online and offline community factors, Brussels: iMinds-VUB-SMIT, 13 August, 32.

individuals differentiate themselves from others, tacitly affirming their uniqueness and distinct individuality to their peers. On the other hand, individuals share common goals and sentiments, which provide them with a social identity [26]. To constitute personal and social identities, it is necessary to manage multiple boundaries and control individual and group information flows. Boundary coordination, however, is challenged by new media, in general, and the properties of OSN, in particular. The properties of media in current digital and networked society have altered dramatically, and they influence everyday life experience: online content persists long after the situation it once originated in, has multiple (invisible) audiences, and is easy to (re)share and find [27, 28]. Moreover, people engage with multiple interconnected networks (e.g., Facebook, Instagram, Snapchat), using multiple artifacts (e.g., smartphones, tablets, different forms of wearable computing technologies) spread over different life spheres, ultimately introducing a new age of living 'in' media, instead of 'with' media [29]. In this study, this age of living in media is labeled with the term 'in-media environment.'

Research Set-up

Over the years, much research has been devoted to the study of boundary coordination in the in-media environment (e.g. [30-34]). Most research is focused on how privacy management manifests itself with a focus on the individual. This study, instead, is focused on how privacy is negotiated and how boundaries are established in youth organizations. The question is: how do youth organizations manage and negotiate their online and offline privacy? 12 focus groups were organized with a total of 78 members (adolescents) and leaders (young adults) of youth organizations to answer this question.

Findings

The results indicate how youth organizations use spatial, temporal, and social boundaries to maintain their online and offline privacy. Youth organizations develop rules about accepting and deleting Facebook friends and what content can be posted, and they sometimes even create a control agent (e.g., 'Facebook police'). Although the context collapse is dealt with by youth organizations, there are two interconnected factors that challenge their boundary coordination: (1) explicit versus implicit privacy rules and (2) individual versus group privacy rules.

First in an online environment, the privacy rules have to be expressed more explicitly. This forces users to actively create boundaries that were not necessary or routinized in an offline environment. For example, in an offline environment, the leaders had difficulties in defining their relationships with older members. OSNs, like Facebook, force users to define relationships by accepting or refusing Facebook friend requests. Even though other means are available to fine-tune those relationships and content sharing (e.g., allowing users access to the main Facebook profile but limiting what content they are able to see), it still requires individuals and groups to think about and define their relationships. Moreover, because privacy rules have to be more explicit, the consequences can also be more prominently present, such as a member being confronted with an ignored friend request from his or her leader. As to the second factor people interact with multiple audiences in even more situations by which they develop layered personal and social identities. In OSNs, a compromise has to be found between individual and group privacy rules, which can be difficult because of the opposite needs of both – that is, the need for a distinct individuality and sharing of common goals and sentiments.

Other studies have found that youngsters can be very creative in managing privacy when not using the privacy settings that they are offered by OSNs (e.g. [28]). It would, therefore, be false to state that not using privacy settings is a sign of not caring about one's privacy. In our study, it was found that group privacy rules are often routinized in the practices of youth organizations, and they often proceed unconsciously, which partially explains the inverse relationship between group privacy rules and perceived privacy control being expressed by the respondents. A similar mistake would be made if youth organizations – or any other type of organization or group, for that matter – that do not express any privacy management strategies were labeled as not caring about how their group information is managed. Therefore, in order to fully understand privacy and boundary coordination in an in-media environment it is useful to differentiate between implicit and explicit privacy rules and take into account the potential tension between different sets of privacy rules.

It is not argued, however, that youth organizations should not actively create boundaries for managing their privacy because many privacy rules are implicit. On the contrary, because of the collapse of boundaries in OSNs, actively formulating certain strategies is not a frivolous luxury either, and helps in coping with privacy turbulences. Moreover, such strategies ensure that organizations translate their notions of privacy into the structure of the platform that is used, rather than the other way around, and it can also help in harmonizing between individual and group privacy rules.

Contemporary western life is media saturated and provides for new and open ways in communicating with each other and constructing social and personal identities. Focusing on the topic of privacy, the results of this study indicate that the adoption of explicit privacy rules and tension between individual and group privacy rules further complicate online boundary coordination. For adequate privacy management in OSNs, it is argued, in line with the recommendations of Wisniewski et al. [34] that service providers should invest in facilitating the negotiation process of content and audiences for both individuals and groups.

3 Conclusion

In order to assess if and to what extent people are 'empowered' in relation to their privacy, different socio-technological aspects were taken into account. For this we have built on insights from studies of digital media and communication in humanities and social science disciplines.

We found that – related to the notion of vulnerability – it is important to understand how technological systems are encoded with regards to privacy (exposure) and then how these systems are decoded by people (coping). The first depends on the interests and context in which a technology comes into existence. The latter depends then on the awareness, attitude, capabilities and practices of people themselves, and how they become configured in their social, cultural and psychological context. Technology is not neutral, but designed to be biased in one way or another. However people can have different readings of these technologies.

The findings also confirm earlier research on how users state that they are concerned about privacy in general [35]. However it has been observed that people's actual behaviour does not correspond to these claims regarding their own privacy [36]. These discrepancies between their claimed attitude towards privacy and their actual privacy behaviour is called the privacy paradox [37]. The research being discussed, stemming from MCS and STS, helps in better understanding and solving the privacy paradox. It gives us a deeper insight into how people engage with social, mobile and ubiquitous media and the possible consequences for managing online privacy.

As to solutions and recommendations this means that we need to extend initiatives based MCS research such as furthering digital literacy practices with educational tools via schools and other organisations, developing appealing Privacy Enhancing Technologies (PET) based on valid user input, setting up clever awareness campaigns, and supporting evidence-based policies and regulations for safeguarding privacy and data protection.

References

1. Freeman, C., Soete, L.: The economics of industrial innovation. Continuum, London (1997)
2. Perez, C.: Technological revolutions and financial capital: the dynamics of bubbles and golden ages. Edward Elgar Pub., Cheltenham (2003)
3. NN: The third great wave. The Economist: Special report - The world economy, pp. 3–4 (2014)
4. van Dijck, J.: The culture of connectivity: a critical history of social media. Oxford University Press, Oxford (2013)
5. van Dijck, J.: Flickr and the culture of connectivity: sharing views, experiences, memories. Memory Studies 4, 1–15 (2010)
6. Pierson, J.: Online privacy in social media: a conceptual exploration of empowerment and vulnerability. Communications & Strategies (Digiworld Economic Journal) 4, 99–120 (2012)
7. Castells, M.: Communication power. Oxford University Press, Oxford (2009)
8. Haythornthwaite, C.A., Wellman, B.: The Internet in everyday life. In: Wellman, B., Haythornthwaite, C.A. (eds.) The Internet in Everyday Life, pp. 3–41. Blackwell, Oxford (2002)
9. Zimmerman, M., Rappaport, J.: Citizen participation, perceived control and psychological empowerment. American Journal of Community Psychology 16, 725–743 (1988)
10. Mansell, R.: From digital divides to digital entitlements in knowledge societies. Current Sociology 50, 407–426 (2002)

11. Villagran, J.C.: Vulnerability: a conceptual and methodological review. United Nations University - Institute for Environment and Human Security (2006)
12. Chambers, R.: Vulnerability, coping and policy (Editorial introduction). IDS Bulletin 37, 33–40 (2006)
13. Ball, K.: Exposure: exploring the subject of surveillance. Information, Communication and Society 12, 639–657 (2009)
14. Watts, M.J., Bohle, H.-G.: The space of vulnerability: the causal structure of hunger and famine. Progress in Human Geography 17, 43–67 (1993)
15. Bohle, H.-G.: Vulnerability Article 1: Vulnerability and Criticality. IHDP Newsletter UPDATE, p. 4. International Human Dimensions Programme on Global Environmental Change Bonn (2001)
16. Norman, D.A.: The psychology of everyday things. Basic Books, New York (1988)
17. Pierson, J., Jacobs, A., Dreessen, K., Van den Broeck, I., Lievens, B., Van den Broeck, W.: Walking the interface: uncovering practices through 'proxy technology assessment'. In: Ethnographic Praxis in Industry Conference - EPIC. American Anthropological Association (2006)
18. Hand, M., Shove, E., Southerton, D.: Explaining Showering: a Discussion of the Material, Conventional, and Temporal Dimensions of Practice. Sociological Research Online 10 (2005)
19. Oudshoorn, N., Pinch, T.J.: How users matter: the co-construction of users and technologies. MIT Press, Cambridge (2003)
20. Akrich, M.: The de-scription of technical objects. In: Bijker, W., Law, J. (eds.) Shaping Technology/Building Society: Studies in Sociotechnical Change, pp. 205–223. MIT Press, Cambridge (1992)
21. Barkhuus, L., Dey, A.: Location-based services for mobile telephony: a study of users' privacy concerns. In: Interact 2003, pp. 709–712. IOS Press (2003)
22. Brandimarte, L., Acquisti, A., Loewenstein, G., Babcock, L.: Privacy concerns and information disclosure: an illusion of control hypothesis. In: iConference. IDEALS, Chapel Hill (2009)
23. Goffman, E.: The presentation of self in everyday life. Penguin, Harmondsworth (1971)
24. Petronio, S.S.: Boundaries of privacy: dialectics of disclosure. State University of New York Press, Albany (2002)
25. Mead, G.H., Morris, C.W.: Mind, self and society from the standpoint of a social behaviorist. University of Chicago Press, Chicago (1934)
26. Hewitt, J.P.: Self and society: a symbolic interactionist social psychology. Allyn and Bacon, Boston (2007)
27. Boyd, D.: It's complicated: the social lives of networked teens. Yale Univesity Press, New Haven (2014)
28. Boyd, D., Marwick, A.: Social privacy in networked publics: teens' attitudes, practices, and strategies. In: Oxford Internet Institute Decade in Internet Time Symposium, p. 22 (2011)
29. Deuze, M.: Media life. Polity, Cambridge (2012)
30. De Wolf, R., Willaert, K., Pierson, J.: Managing privacy boundaries together: exploring individual and group privacy management strategies in Facebook. Computers in Human Behavior 35, 444–454 (2014)
31. Lampinen, A., Lehtinen, V., Lehmuskallio, A., Tamminen, S.: We're in it together: Interpersonal management of disclosure in social network services. In: SIGCHI Conference on Human Factors in Computing Systems, pp. 3217–3226. ACM (2011)

32. Litt, E.: Understanding social network site users' privacy tool use. Computers in Human Behavior 29, 1649–1656 (2013)
33. Stutzman, F.D., Hartzog, W.: Boundary regulation in social media. In: ACM Conference on Computer Supported Collaborative Work and Social Computing, pp. 769–778. ACM (2009)
34. Wisniewski, P., Lipford, H., Wilson, D.: Fighting for my space: Coping mechanisms for SNS boundary regulation. In: SIGCHI Conference on Human Factors in Computing Systems, pp. 609–618. ACM (2012)
35. Pötzsch, S.: Privacy awareness: A means to solve the privacy paradox? In: Matyáš, V., Fischer-Hübner, S., Cvrček, D., Švenda, P. (eds.) The Future of Identity. IFIP AICT, vol. 298, pp. 226–236. Springer, Heidelberg (2009)
36. Deuker, A.: Addressing the privacy paradox by expanded privacy awareness – the example of context-aware services. In: Bezzi, M., Duquenoy, P., Fischer-Hübner, S., Hansen, M., Zhang, G. (eds.) Privacy and Identity. IFIP AICT, vol. 320, pp. 275–283. Springer, Heidelberg (2010)
37. Barnes, S.B.: A privacy paradox: social networking in the United States. First Monday 11 (2006)

Engaging with Charities on Social Media: Comparing Interaction on Facebook and Twitter

Christopher Phethean$^{(\boxtimes)}$, Thanassis Tiropanis, and Lisa Harris

Web Science Institute, University of Southampton, Southampton, UK
{C.J.Phethean,T.Tiropanis,L.J.Harris}@soton.ac.uk

Abstract. Social media are commonly assumed to provide fruitful online communities for organisations, whereby the brand and supporter-base engage in productive, two-way conversations. For charities, this provides a unique opportunity to reach an audience for a relatively low cost, yet some remain hesitant to fully embrace these services without knowing exactly what they will receive in return. This paper reports on a study that seeks to determine the extent to which these conversations occur, and compares this phenomenon on Facebook and Twitter for a sample of UK-based charities. Focus was placed on analysing conversations as signs of developing relationships, which have previously been shown to be a key target for charities on social media. The results of this study find that while there is an expected proportion of the audience who prefer to listen rather than engage, there is strong evidence of a core group of supporters on each site who repeatedly engage. Interestingly, disparities between how this occurs on Facebook and Twitter emerge, with the results suggesting that Facebook receives more conversations in response to the charities' own posts, whereas on Twitter there is a larger observable element of unsolicited messages of people talking about the charity, which in turn produces a differing opportunity for the charity to extract value from the network. It is also found that posts containing pictures receive the highest number of responses on each site. These were a lot less common on Twitter and could therefore offer an avenue for charities to increase the frequency of responses they achieve.

Keywords: Social media · Charities · Marketing · Communication

1 Introduction

Social media's popularity in the contemporary world makes it easy to think that people interacting and engaging online can transform the process of information production and dissemination and produce a constant stream of brand advocates, supporters and critics for any organisation. In an idealistic sense, analysing this at scale would allow the organisation to determine the current perception of their services, which they could then use to encourage further engagement and generate a more loyal supporter base who help to contribute to their own marketing efforts. For charities, it is essential that they know whether or not this is really possible when they are making the decision on whether to

© Springer International Publishing Switzerland 2015
T. Tiropanis et al. (Eds.): INSCI 2015, LNCS 9089, pp. 15–29, 2015.
DOI: 10.1007/978-3-319-18609-2_2

allocate limited funds to developing and maintaining a social presence online. A problem arises, however, in that the level of understanding around how effective these services actually are for establishing these social relationships is limited. It is easy to assume, given the constant media hype around them, that these sites breed user engagement and citizen activism. However, there are also established 'rules' to suggest that the percentage of any community online which actually engages is small [10]. If this is the case for a charity's online presence, then this could have great implications on their perception of the 'value' or 'worthiness' of social media to them as an organisation.

This paper, therefore, seeks to assess what evidence there is on social media of developing relationships between charity and supporter. This is motivated by previous work interviewing members of social media teams at various charities in the UK which found that relationship building was seen as one of the key aims and values of using social media [13]. Therefore, the goal of this paper is to show to what extent this phenomenon is actually apparent. Further contributions are made by comparing two of the largest social networks—Facebook and Twitter— to examine whether there is any difference in the levels by which people appear to engage and how these each contribute to the information transformation offered by the Internet. Studies into social media frequently look at Twitter due to the ease of use of the API to collect data, yet few compare this phenomenon to what occurs on other networks. While some (e.g. [11], [15]) do explore this difference from a brand perspective, it remains an under-researched area for a topic with as much interest as social media.

Most significantly, however, this paper emphasises that statistics measuring social media interactions which may represent a developing relationship can only reveal so much about this area, and that a more holistic approach which moves away from looking solely at social media statistics is required. A qualitative analysis of some of the most engaging messages is provided to investigate the content and themes that lead to the most signs of interaction. Future directions are outlined that insist on the integration of quantitative social media statistics, qualitative social media content analysis and the views, goals and needs of the charity itself.

2   Background

2.1   Relationship Building by Charities on Social Media

For both corporations and nonprofit organisations such as charities, social media can offer an opportunity to produce a number of different outcomes. Spreading awareness of a new product or campaign, referring users to websites in order to increase traffic, generating buzz to gain media attention, and building relationships with audience members are all possibilities. Additionally each social media site offers a unique set of features that affords different types of interaction and which may make certain aims more suitable for particular sites. However, there is an equal sense that sometimes use of these sites maybe does not match their perceived value. Twitter, for example, is often presented as a great way to offer rapid customer service and to interact with supporters through two-way engagement. A content analysis of USA-based nonprofits' tweets instead indicates that their focus is more on sending one-way messages in order to broadcast information [19], and similar studies have also suggested a reluctance to move away from primarily information spreading behaviour [9].

A case study of the American Red Cross elicits aims for social media including the discovery of public perception, highlighting areas of improvement and generating media attention [3]. It is claimed by the authors of the study that the American Red Cross can be used as a model to follow for organisations wanting to effectively utilise social media, with an interactive two-way communicative approach essential [3]. Familiar benefits of this were listed: rapid community service and the elicitation of positive and negative feedback [3]. It must be questioned, however, how representative this feedback is if it is only sourced from a small subset of the audience who are responding. Nonetheless, interactivity on social media is said to be essential in allowing productive relationships to develop with supporters; a lack of it could potentially turn supporters away [18], and it can generally increase trust [8]. Consequently, even if the relative proportion of engaging users is low, just by showing that there is a two-way, responsive conversation could increase the trust of those users who do not with to interact online.

Interviewing members of charities who were involved in their social media presence, the current authors identified recurring themes about why social media was used, and what they hoped to get out of using it [13]. One of the most important reasons for using social media was to develop relationships with supporters, and achieving 'action' through donations was seen as a positive side-effect of doing this rather than a primary aim. There was a slight favouring towards Facebook for achieving this, especially as it provided a centralised location for people to provide support and advice—both from the charity to supporter, and supporter-to-supporter [13]. There was less clarity regarding the actual success of these sites in achieving relationship building. Accounts of favourable outcomes were given relating to the number of 'likes' content received on Facebook, which does not necessarily indicate that the users involved have a strong relationship with the charity [13].

2.2   Online Listeners and Slacktivists

Within any community - online or off - there will be a portion of it that does not interact or contribute, but frequently consumes the content created by others. Throughout the 1990s and early 2000s, it began to be established that for an online community, around 90% of members would fall in to this category - often labelled as 'lurkers' (for example, see [10]). [16] claims that there is a difference between "passive lurkers" and "active lurkers' who go on to use the information gathered from the online community in an offline setting, providing the beginning of an argument to suggest that the common negative perception towards these users may be undeserved. More recently, [6] discusses the stigma attached with the term 'lurker' and suggests a reconceptualisation to 'listener' instead - reflecting an equally important role in any online conversation as that of speaking.

This also more accurately reflects the activities that many of these users will be carrying out but while a listener can be engaged in a conversation just as much as a frequent commenter, they may be missed by social media analyses looking solely for interactions. Furthermore, research on Facebook has shown that people frequently underestimate the size of the audience that is exposed to a post, as inaccurate measures such as amounts of feedback and friend counts are used, which do not reflect how many people actually listened [1].

What may be possibly more misleading when examining social media for these interactions is the phenomenon of 'slacktivism' - described in [14] as activities that are low cost and risk, and generate satisfaction in the actor. If this satisfaction is generated to replace that gained by actually doing something - rather than just clicking 'like', for example - then these actions can be misinterpreted by charities whose returns will not correlate with the actions occurring on social media.

2.3   Conversations as Indications of Relationships?

While actions that just require 'clicking' to complete - such as liking on Facebook and retweeting on Twitter - may be looked at as being possible instances of slacktivism, other actions on these sites can reflect more meaningful signs of engagement, and may indicate a stronger relationship between charity and supporter. In earlier work by the current authors, a preliminary framework for social media measurement was presented (focusing at the time on Twitter), with retweets and replies listed as indications of engagement [12]. Previous work has provided methods of analysing retweets as a mechanism for communication and disseminating topics through the network ([2, 17]). However, while these papers show the reasons behind retweeting are varied, the replies metric appears to be a more appropriate measure for discovering strong, developed relationships: a reply requires a larger investment in terms of time and effort than simply clicking retweet, and as such [4] describes textual comments on Facebook as the highest possible level of engagement. In addition, to facilitate the two-way communication discussed above on Twitter, replies - and posts by the audience mentioning the charity that can be replied to - are essential aspects that must be analysed. Likewise, on Facebook, the reply or comment feature would showcase more than a like, and represent evidence of a strengthened relationship. In [7], the finding that 'requests and suggestions', 'expressing affect' and 'sharing' are popular intentions for participation, also suggests that textual comments will play a key role in users' interactions. In addition, overcoming users opting for the 'safest' options - those that do not provoke reactions from other members of the community, e.g. liking - should be encouraged and that improving the level of activity through conversational interactions should be sought [7]. This again suggests a significant value in conversations on social media, as they go beyond the 'easy' and 'safe' options, to show the organisation that there really is a valuable relationship present.

For the purposes of this paper, replies and mentions on each site will be used as a representation of engagement. While retweets, likes and shares are not being focused on, the authors do not disregard their impact - indeed it is appreciated that each of these mechanisms plays a vital role in the social media marketing mix. In the current study, however, the evidence for strong relationships is sought based on the discussion above about this being a key aim for charities on social media, and that replies seem to be a viable channel to indicate this. The current study seeks to answer the following questions:

RQ1. Does either Twitter or Facebook show evidence of more sustained relationships between supporters and charity?

RQ2. Do posts by a charity on either Twitter or Facebook tend to generate more engagement than on the other?

RQ3. Do certain types of posts by a charity on either site tend to generate more or longer conversations with the audience?

3   Methodology

Based on the research questions devised above, a number of hypotheses are proposed. For RQ1, examining whether Twitter or Facebook creates more sustained relationships, it will be necessary to identify commenting users as discussed above to signify users who have a more developed relationship, and to determine how developed those relationships are through the intensity by which they engage. To further explore this area, and to indicate whether achieving engagement on one site could be down to the strategies carried out on that site, rather than an underlying popularity and tendency for users to engage anyway, the correlation between the scores on the two sites will be examined. This leads to the first two hypotheses:

H1a. Engaging supporters for each charity will post significantly more comments on one of Facebook or Twitter than the other.

H1b. If engagement is related to the overall popularity of the charity, then charities with more engaged supporters on Facebook will also have more engaged supporters on Twitter.

After testing these two hypotheses, it will be important to relate this back to the authors' previous work, which suggested that charities believed Facebook is the better platform for developing relationships [13].

Looking at RQ2, it can be seen that providing insights in to this question will help reinforce this argument, and go on to provide recommendations to charities as to which site would be worth spending more time on—or which would be better to dedicate limited resources to. As with the hypotheses for RQ1, it is expected that from the opinions shared in [13], there will be noticeable differences in the ways in which users respond to posts on each of the two networks. Whereas H1a and H1b looked at overall engagement towards the charity, RQ2 focuses on the responses to the charities' posts in order to discover the importance of the charity showing an interactive presence on the site, which emanates from the discussion above about this being a vital part of building relationships [8, 18]. The following hypotheses are proposed, focusing on the conversational responses to posts by an audience:

Table 1. Data overview

Charity	Facebook Data			Twitter Data		
	Likes	Charity's Root Posts	Commenters	Followers	Charity's Root Tweets	Mentioners
DUK	57834	135	3012	76808	820	11472
DT	583569	451	12787	115687	1249	21083
WH	529	76	27	586	424	179
HfH	324490	269	7371	224259	2901	82154
JfG	4973	252	495	12045	3818	17104
NT	144701	311	8144	210241	763	27040
WT	40067	142	2941	48449	1227	7890

H2a. There is a significant difference between the number of comments per supporter on charities' posts on Facebook and Twitter.

H2b. There is a significant difference between the number of supporters who comment on the charities' posts on Facebook and Twitter.

H2c. If conversations between the charity and an engaged audience are occurring on each of the two sites, then on days when the charity posts more messages, the audience will also post more messages on the same site.

Finally, in order to investigate RQ3, based on a finding in [13] that suggested charities thought posts containing pictures were particularly 'successful', it is hypothesised that:

H3a. On each site, posts containing pictures will on average produce more comments than any other post format.

By investigating H3a, it is hoped to be able to suggest strategies that will work most effectively for charities that wish to develop engagement and relationships with their supporters.

3.1   Dataset

A sample of 7 charities was used for this study, the 5 from the authors' previous study [13] (Diabetes UK, The Dogs Trust, Help for Heroes, Jeans for Genes, The Woodland Trust), along with two more: The National Trust and Wessex Heartbeat. The sample ensured that charities of various sizes (regarding their income) were chosen, and 6 of the 7 charities have been interviewed to ensure their views towards social media and their intended uses for it are known beforehand. For each charity, a dataset of 6-months worth of data was collected for each site:

from Twitter, a variation of the University of Southampton Tweet Harvester[1] was used to collect tweets over the course of the study, whereas the Facebook dataset was collected retrospectively using a combination of the Facebook Graph API and Facebook FQL. For both sites, the collected data covered the period June–December 2013. The Twitter dataset consisted of any tweet sent by the charity, to the charity, or mentioning the charity, while from Facebook every post and associated comment made on the charity's page was collected (including posts by a supporter directly on to the page and their resulting comments). For each conversation the root post ID, the root post format, conversation chain length, number of conversation participants, whether or not the charity started the conversation, and whether or not the charity replied in the conversation were recored. While this was trivial for analysing responses to Facebook posts by simply collecting and analysing the list of comments, the process was more complex for Twitter and as such an algorithm based on the work in [5] was produced to form tweets into conversation chains.

Additionally, each user who participated in a conversation was recorded along with how many posts they made in the 6 month period, and how many conversations these fell in to. In addition, for each charity the follower count (on Twitter, as of 5th February 2014) and the number of page likes (for Facebook, as of 4th February 2014) were collected to enable proportional calculations to be carried out. A summary of the dataset is provided in Table 1. In total, 493328 posts (root posts, comments and replies) were analysed, from 201699 users.

The top 5 posts sorted by number of comments from each network for each charity were extracted so that qualitative content analysis could be carried out in order to determine whether there were any charity-specific or overall themes that appeared to cause the highest levels of conversation. This would assist in making any recommendations to the charities about what type of content produces the most desirable results, and is essential in order to contextualise the statistical work that will be used to assess the hypotheses above.

4   Results

4.1   Commenter Analysis

For RQ1, it was necessary to examine the behaviours of commenters on both sites towards each charity. Calculations were made to assess how many posts each user made, with the results shown in Table 2. A Wilcoxon signed-rank test was then carried out on these values to determine whether one site produced significantly higher values. With Twitter observably higher in each case, the test indicated that there was a difference ($z=-2.366$, $p<0.05$, $r=-0.63$). To examine this area further, the number of commenters who posted more than once (repeated engagers) and 6 or more times (once per month) were calculated. These values are displayed in Table 3. Using the monthly observers values (and acknowledging that this does not necessarily mean one post per month was made—a single

---

[1] http://tweets.soton.ac.uk

Table 2. Commenter Statistics

Charity	Average Posts Per Commenter (FB)	Average Posts Per Commenter (TW)
DUK	2.66904	2.98087
DT	2.27573	2.33973
WH	1.82853	4.90465
HfH	1.65227	2.04518
JfG	1.72773	3.19667
NT	2.51088	3.19667
WT	1.86335	2.26464

conversation of 6 posts could also place a user into this category), a Wilcoxon signed-rank test was again carried out showing that there is a difference between the two sites in favour of Twitter ($z=-2.197$, $p<0.05$, $r=-0.59$). For repeat engagers the same calculation showed that there was no significant difference between the sites ($z=-1.352$, $p>0.05$, $r=-0.36$). This section of analysis therefore indicates that per interacting supporter, more interactions are made on Twitter than Facebook, which therefore supports H1a.

An 'engagement index' was then made for each charity on each site. On Facebook this was the average of Z-scores for the number of commenters per likes ($M=0.055$, $SD=0.026$), the average number of conversations each user participated in ($M=1.776$, $SD=0.258$) and the average number of posts made by each user per conversation ($M=1.163$, $SD=0.084$). On Twitter these were the number of posters per followers ($M=0.388$, $SD=0.464$), the average number of conversations ($M=2.668$, $SD=0.818$) and the average number of tweets made by each user in each conversation ($M=1.062$, $SD=0.050$). Creating Z-scores for each charity's own score in relation to these, and averaging them provided an index for each site. To test whether charities with more engaged users on one site also had more engaged users on the other, a Spearman correlation was calculated on these values, and showed no significant correlation ($r=0.321$, $p=0.482$), therefore H1b was not supported.

4.2 Audience Response Analysis

RQ2 focused more on how the audience responded to the charities' posts on each network. Looking at the data from the perspective of the *posts*, rather than the posters, calculations were made to find the number of comments or replies per charity-authored post on the two networks. Again, a Wilcoxon signed-rank tests were carried out on the results. Firstly on the average number of comments per post per like (or follower), which this time showed Facebook as being consistently higher ($z=-2.366$, $p<0.05$, $r=-0.63$), meaning that per supporter on each site, Facebook produced a higher number of comments or replies on each of the charities's posts than Twitter, and supported H2a. This was shown again when

Table 3. Repeated Engager Statistics (out of supporters who have commented at least once)

Charity	Repeat Engagers		Monthly Engagers	
	FB	**TW**	**FB**	**TW**
DUK	1159 (38.5%)	4134 (36%)	207 (6.9%)	1000 (8.7%)
DT	5091 (32.2%)	6182 (29.3%)	884 (5.6%)	1152 (5.5%)
WH	10 (37.0%)	59 (33%)	2 (7.4%)	14 (7.8%)
HfH	1894 (25.7%)	16922 (20.6%)	180 (2.4%)	3308 (4%)
JfG	130 (26.3%)	4637 (27.1%)	10 (2%)	769 (4.5%)
NT	2896 (35.6%)	9904 (36.6%)	545 (6.7%)	2205 (8.2%)
WT	916 (31.1%)	2462 (31.2%)	129 (4.4%)	427 (5.4%)

Table 4. Post Replies Statistics (out of all supporters (likes or followers))

Charity	Average Responses Per Post Per Supporter		Average Responders Per Post Per Supporter	
	FB	**TW**	**FB**	**TW**
DUK	0.00055	0.0005	0.00042	0.00002
DT	0.00010	0.00001	0.00009	0.00001
WH	0.00077	0.00029	0.00062	0.00021
HfH	0.00010	0.00000	0.00009	0.00000
JfG	0.00050	0.00002	0.00042	0.00001
NT	0.00033	0.00003	0.00022	0.00001
WT	0.00138	0.00002	0.00126	0.00002

looking at the average number of *commenters* or posters per like or follower on each site, with Facebook again consistently higher ($z=-2.366$, $p<0.05$, $r=-0.63$). This data is summarised in Table 4 and provides an indication that Facebook provides a higher proportion of interacting or engaged supporters than Twitter, supporting H2b.

Finally, to test H2c, a Spearman correlation was calculated as a rudimentary analysis of the timestamps of charity posts and audience posts. For each day in the 6 month study period, a tally of how many posts were made by charity and audience was gathered. Table 5 shows that for each charity there is a significant correlation between when the charity itself posts, and when their audience posts. While H2c is supported, it is clear that the strengths of the correlations are varied and many comments may well be being made in an unsolicited way. When run through the Wilcoxon signed-rank test, these correlations showed that there was no significant difference between the strengths in correlation between each site ($z=-0.338$, $p>0.05$, $r=-0.09$).

Table 5. Charity and Supporter Post Date Correlation

Charity	FB Correlation	TW Correlation
DUK	r=0.643 p<0.01	r=0.6 p<0.01
DT	r=0.413 p<0.01	r=0.718 p<0.01
WH	r=0.341 p<0.01	r=0.269 p<0.01
HfH	r=0.522 p<0.01	r=0.302 p<0.01
JfG	r=0.54 p<0.01	r=0.662 p<0.01
NT	r=0.541 p<0.01	r=0.53 p<0.01
WT	r=0.666 p<0.01	r=0.621 p<0.01

Table 6. Post Format Responses (Facebook)

Format	Comments/Posts	Commenters/Posts	ResponseIndex
Statuses	23.68	19.71	-0.289
Videos	24.43	21.69	-0.163
Pictures	43.74	37.62	1.406
Links	15.36	13.07	-0.954

4.3   Post Format Analysis

The quantities of each different post type (statuses, pictures, videos and links) sent by the charity were counted on the two sites. For each, the average number of comments per post (FB: M=26.806, SD=12.012, TW: M=1.496, SD=0.873) and the average number of commenters per post (FB: M=23.020, SD=10.406, TW: M=0.966, SD=0.571) were calculated (across the entire dataset for each site), with Z-scores calculated on each. An index for the engagement with each post format was then calculated as the average of these two values (Table 6 and Table 7). On Twitter, no posts were returned labelled as containing video content, so this row was excluded from the calculation. Posts containing pictures were shown to be the most engaging from a conversational point of view on both sites, receiving far higher amounts of commenters, from more commenters, than any other format. Particularly noteworthy was that on Facebook, a picture posted by a charity received on average over 37 unique commenters—on Twitter this was only 1.56 commenters per post. These results help to support H3a, and can be used to provide a strong recommendation to charities looking to increase the amount of comments that they receive.

4.4   Top Post Content Analysis

From looking qualitatively at the content of the messages that received the highest number of comments, several recurring types of message were discovered.

Table 7. Post Format Responses (Twitter)

Format	Replies/Posts	Repliers/Posts	ResponseIndex
Statuses	0.64	0.423	-0.966
Videos	N/A	N/A	N/A
Pictures	2.38	1.56	1.030
Links	1.46	0.91	-0.065

On Facebook, posts asking an informal question to the fans of the charity's page were common (12/35 posts), as were those promoting a competition (10/35). 7 of the posts directly referred to pictures included in them, and 21 of the 35 were classed as being informational messages for a variety of purposes (such as linking to relevant content or reporting media attention). Informally toned messages (10/35) were more common than any formal or authoritative messages (combined 4/37). On Twitter, informal questions were again a popular type of content to be commented upon (16/35), and informational messages (14/35) also tended to be replied to. Promoting a competition (4/35) in messages were less commonly popular than on Facebook, but general messages of thanks (4/35) indicated that these types of tweet gained a fair amount of replies. In two cases on both Facebook and Twitter, informal questions made up the majority (5/5, 5/5, and 4/5, 4/5) of all the charities' most popular tweets. There are indications here that informal questions are particularly effective at generating responses from the audience.

5 Discussion

It is interesting to discover that for the sample of charities in this study, Twitter appeared to accomodate supporters who made more interactions each, compared to Facebook (H1a). Yet when looking at the data from the point of view of responses to the charities' own posts and in relation to the number of likes or followers each charity possessed, Facebook posts received more comments (H2a), and more commenters (H2b) than Twitter. It is possible that the disparity arises from H1a and could be down to the nature by which the data was collected: while the Twitter Search API allowed any messages mentioning the charity to be gathered, the Facebook data collection was restricted to what appeared on their page only, and so reflects the observable interacting users (as H2a and H2b focused on responses to the charities' posts, this is not an issue). However, with the publicity of conversations such as these one of the major advantages of social media, this suggests a great opportunity on Twitter to discover these unsolicited discussions and for the charity to then take advantage.

There is clearly a significant amount of discussion about charities occurring on Twitter, so why are the values in Tables 4, 6 and 7 so much smaller than for

Facebook? Returning to the literature discussed earlier, it is possible that this is symptomatic of the 'wrong' types of messages being sent by the charities—messages that are not conducive to conversation, and are one-way broadcast messages instead [19], [9]. With pictures appearing to be the post format that produces the most comments in response (H3a), it is interesting to note that pictures accounted for nearly 76% of the charities' posts on Facebook, while on Twitter it was a little under 14%.

Alternatively, the difference in post response rate could be down to the fact that charities do not see Twitter as a channel for relationship building in the way that they do with Facebook, supporting the views presented in the current authors' previous study [13]. It is important to note that this does currently appear to be the case, and their actual use does correlate with their perceived intentions and beliefs. If charities do see Twitter as more of a mechanism for promoting awareness and spreading information—as claimed in [13]—then signs of conversation in response to their posts would be less apparent. However it appears from the qualitative aspects of this study that there is some evidence to suggest that tweets attempting to elicit a reaction—primarily asking informal questions—are still the most popular on Twitter in terms of replies received, and engagement in this way is still possible. However a more in-depth examination of content-type and popularity is required to verify this further.

At this point, another interesting question arises. What signifies a stronger relationship: a supporter repeatedly replying to a charity's post, or a supporter regularly posting messages without a prompt from the charity itself? This is a key question to move forward in this area, requiring further study of the types of message being sent. Something is occurring which statistics are failing to account for, suggesting more in-depth qualitative methods are required now that a general understanding of the area has been obtained. The suggestion, however, is that there is a disparity of value between Facebook and Twitter—while Facebook may by better for developing relationships with continually interested supporters, Twitter's value may be in harbouring unsolicited mentions of what a wider range of people choose to do for the charity, which the charity itself can utilise for promotion. A more detailed time-series analysis of this area could provide additional understanding of this relationship.

Insights are gained from the unsupported hypothesis, H1b, which stated that charities with more engaged supporters on one site would have more engaged supporters on the other. The correlation showed that this was not supported, and suggests that engaging users on one site may not be symptomatic of a pre-engaged and more active audience—if this was so then both sites would tend to receive high levels of engagement compared to charities that did not. Instead, it suggests that something the charity is doing on one of the sites is probably 'working' more than on the other to stimulate conversation—again this comes down to what they look to get out of each, and requires further research to establish the overall state of a community spanning multiple social networks.

5.1   Limitations and Summary of Contribution

This paper examined social media interactions around 7 UK charities on Facebook and Twitter. This is a small sample and caution must be taken before inferring the indications to the wider population. There was also limited qualitative analysis carried out which could provide much richer insights in to what people are actually conversing about. Only textual interactions were analysed, and this was only examined from the perspective of relationship building—there are other aims that are also important to charities.

The findings of this study suggest that the ways in which supporters engage with charities differs between sites, supporting the perceptions of the charities themselves presented in [13]. There is evidence to back proposals that Facebook could be better for posting messages to encourage a known supporter-base to respond, whereas Twitter appears to be a more valuable for discovering unsolicited mentions and accounts of support from any users on the network. Whether this is the cause of, or effect of, the results of studies such as [19] that claim charities in the USA focus on sending one-way messages rather than encouraging conversation is yet to be discovered. The qualitative part of this study showed that informal questions were regularly the most replied to type of tweet, and this gives some indication that the claim of 'wrong use' can not be entirely supported.

While the proportions of commenters compared to the number of supporters is low, the proportion of users posting more than once in 6 months is encouraging. It is to be expected, given the literature review, that the majority of the community will be content to listen. These results help to highlight how valuable small portions of an audience are in creating a large amount of conversation around a topic—and the techniques used to locate these users can now be taken further to assess their overall contribution to the buzz around the charity.

Based on the discoveries in this paper, we can make some suggestions to charities wishing to develop their relationship building strategies on social media. Facebook appears to be the more suitable choice for generating discussions about relevant topics with the dedicated, committed supporter base. Twitter seems to hold great potential value for finding the extra, satellite discussions about the charities, and then supporting and amplifying these using Twitter's ability to quickly disseminate and spread messages. On both sites, pictures should be used when the charity desires a conversation or many replies. It seems apparent, however, that a statistical approach such as this to analysing social media can only say so much, despite providing initial insights and an overall picture of this area. Further qualitative analysis of messages is required, and this must be used in conjunction with the aims and view of the charities discovered in [13] to determine further where the value in social media truly arises from.

Acknowledgement. This research was funded by the Research Councils UK Digital Economy Programme, Web Science Doctoral Training Centre, University of Southampton. EP/G036926/1.

References

1. Bernstein, M.S., Bakshy, E., Burke, M., Karrer, B.: Quantifying the invisible audience in social networks. In: Proceedings of the SIGCHI Conference on Human Factors in Computing Systems, CHI 2013, pp. 21–30. ACM, New York (2013), http://doi.acm.org/10.1145/2470654.2470658
2. Boyd, D., Golder, S., Lotan, G.: Tweet, tweet, retweet: Conversational aspects of retweeting on twitter. In: 2010 43rd Hawaii International Conference on System Sciences (HICSS), pp. 1–10 (2010)
3. Briones, R.L., Kuch, B., Liu, B.F., Jin, Y.: Keeping up with the digital age: How the american red cross uses social media to build relationships. Public Relations Review 37(1), 37–43 (2011), http://www.sciencedirect.com/science/article/pii/S0363811110001335
4. Cho, M., Schweickart, T., Haase, A.: Public engagement with nonprofit organizations on facebook. Public Relations Review 40(3), 565–567 (2014), http://www.sciencedirect.com/science/article/pii/S0363811114000241
5. Cogan, P., Andrews, M., Bradonjic, M., Kennedy, W.S., Sala, A., Tucci, G.: Reconstruction and analysis of twitter conversation graphs. In: Proceedings of the First ACM International Workshop on Hot Topics on Interdisciplinary Social Networks Research, HotSocial 2012, pp. 25–31. ACM, New York (2012), http://doi.acm.org/10.1145/2392622.2392626
6. Crawford, K.: Following you: Disciplines of listening in social media. Continuum 23(4), 525–535 (2009), http://www.tandfonline.com/doi/abs/10.1080/10304310903003270
7. Cvijikj, I.P., Michahelles, F.: Understanding the user generated content and interactions on a facebook brand page. International Journal of Social and Humanistic Computing 2(1), 118–140 (2013), http://dx.doi.org/10.1504/IJSHC.2013.053270
8. Jo, S., Kim, Y.: The effect of web characteristics on relationship building. Journal of Public Relations Research 15(3), 199–223 (2003), http://www.tandfonline.com/doi/abs/10.1207/S1532754XJPRR1503_1
9. Lovejoy, K., Saxton, G.D.: Information, community, and action: How nonprofit organizations use social media. Journal of Computer-Mediated Communication 17(3), 337–353 (2012), http://onlinelibrary.wiley.com/doi/10.1111/j.1083-6101.2012.01576.x/abstract
10. Nielsen, J.: Participation inequality: Encouraging more users to contribute (October 2006), http://www.useit.com/alertbox/participation_inequality.html (last accessed: November 26, 2014)
11. Paek, H.J., Hove, T., Jung, Y., Cole, R.T.: Engagement across three social media platforms: An exploratory study of a cause-related PR campaign. Public Relations Review 39(5), 526–533 (2013), http://www.sciencedirect.com/science/article/pii/S0363811113001379
12. Phethean, C., Tiropanis, T., Harris, L.: Measuring the performance of social media marketing in the charitable domain. In: Web Science 2012, Evanston, IL, USA (June 2012)
13. Phethean, C., Tiropanis, T., Harris, L.: Rethinking measurements of social media use by charities: A mixed methods approach. In: Web Science 2013. ACM, Paris (2013)
14. Rotman, D., Vieweg, S., Yardi, S., Chi, E., Preece, J., Shneiderman, B., Pirolli, P., Glaisyer, T.: From slacktivism to activism: participatory culture in the age of social media. In: Conference on Human Factors in Computing Systems. ACM, Vancouver (2011), http://dl.acm.org/citation.cfm?id=1979543

15. Smith, A.N., Fischer, E., Yongjian, C.: How does brand-related user-generated content differ across YouTube, Facebook, and Twitter? Journal of Interactive Marketing 26(2), 102–113 (2012), http://www.sciencedirect.com/science/article/pii/S1094996812000059

16. Takahashi, M., Fujimoto, M., Yamasaki, N.: The active lurker: influence of an in-house online community on its outside environment. In: GROUP 2003 Proceedings of the 2003 International ACM SIGGROUP Conference on Supporting Group Work. ACM Press (2003), http://portal.acm.org/citation.cfm?doid=958160.958162

17. Tinati, R., Carr, L., Hall, W., Bentwood, J.: Identifying communicator roles in twitter. In: Proceedings of the 21st International Conference Companion on World Wide Web, WWW 2012 Companion, pp. 1161–1168. ACM, New York (2012), http://doi.acm.org/10.1145/2187980.2188256

18. Waters, R.D., Burnett, E., Lamm, A., Lucas, J.: Engaging stakeholders through social networking: How nonprofit organizations are using facebook. Public Relations Review 35(2), 102–106 (2009), http://linkinghub.elsevier.com/retrieve/pii/S0363811109000046

19. Waters, R.D., Jamal, J.Y.: Tweet, tweet, tweet: A content analysis of nonprofit organizations' twitter updates. Public Relations Review 37(3), 321–324 (2011), http://www.sciencedirect.com/science/article/pii/S0363811111000361

Empowering Female-Owned SMEs with ICT in A Group of Selected Arab Countries and Brazil

Mona Farid Badran[✉]

Faculty of Economics and Political Science,
Cairo University, Giza, Egypt
samifarah.mona@gmail.com

Abstract. This research paper embarks on a comparative empirical study to investigate the impact that ICT plays on empowering women entrepreneurs in 5 developing/emerging countries, namely Egypt, Jordan, Morocco, Algeria, as a group of Arab countries and Brazil. The World Bank's Investment Climate Assessment Surveys (ICA) is the primary source of data for the four Arab countries and Brazil. The ICA database provides comparable enterprise level data based on similar sampling techniques. The results obtained from the empirical study reveal that in the selected Arab countries, the increase in female owned SMEs is associated with a decrease in the Internal Rate of Return. However, when we control for ICT in terms of ICT index constructed using the Principal Component Analysis technique (PCA), the female owned SMEs becomes statistically insignificant; this is also the case with the ICT index. This implies that IRR is negatively associated with the female owners of the SME, and there is a no association between IRR and the access and use of ICT. In Brazil, however, neither gender nor ICT plays any role in the profitability of SMEs. However, as for the other measure for economic performance, namely the labor intensity, the findings reveal that in the selected Arab countries, the ICT index has a positive, statically significant, association with labor-intensity, while in Brazil the usage of a Website has a negative, statistically significant, association with the labor-intensity.

Keywords: SMEs · Females · Arab countries · Brazil · ICT · ICA survey
JEL-classification: J16 · M13

1   Introduction

Empowering women in developing countries, including the Arab countries, as well as in emerging countries such as Brazil, is considered a vital issue for social and economic development [22] [17]. Female entrepreneurship represents a potentially valuable tool for promoting growth and elevating poverty and combating gender inequality[14]. In this respect, it is important to refer to the Grameen Bank model and other micro-finance schemes that lend mainly to women, to buy cell-phones and provide mobile pay phone services [21][18]. It is an established fact that one of the UN Millennium Development Goals is gender equality and women empowerment,

© Springer International Publishing Switzerland 2015
T. Tiropanis et al. (Eds.): INSCI 2015, LNCS 9089, pp. 30–48, 2015.
DOI: 10.1007/978-3-319-18609-2_3

and it has been argued that ICT can be the vehicle to achieve this goal[2]. In the realm of SMEs in the considered countries, female-owned firms represent 30-37 percent of all SMEs. Furthermore, jobs in small and medium enterprises (SMEs) account for more than half of all formal employment worldwide [13].

Historically, starting in the 1990, many SMEs in developing countries, albeit a smaller number, began to include modern information and communication technologies in their enterprises. This trend can result in an increase in the returns for enterprises and an increase in their productivity. In addition, it can make training and education more accessible for workers. This can eventually lead to increase of employability of low skilled workers [13].

Another vital motivation of the present study is the fact that female-owned SMEs can, by using the Information and Communication Technologies (ICT), move up the value chain and reach higher value–added products. Thus, owners of SMEs reach out to new technologies and innovations, while the latter require more skilled workers [13].

This progress can also result in reduction of poverty, especially since the majority of female owned SMEs are entering the business because of necessity rather than opportunit[7].

In some cases, business owners and managers, regardless of their gender, lack the requirement to manage and promote their businesses using new technologies. This results in limiting their potential for growth and job creation. Innovation and technology advances can be regarded as a solution for the obstacles facing SMEs, by promoting the skills of the labor force in general and the female owners of SMEs in particular [13]. Women entrepreneurs are offered new opportunities by ICTs to start and grow their businesses. By using traditional and new ICTs, female–owned SMEs are connecting with their clients, becoming more productive and innovative. Thus, without access to these, entrepreneurs are not totally exploiting the business opportunity in their market. First and foremost, thanks to the mobile revolution and the new concept of ubiquitous Internet access, access to ICT tools is now much easier even to the rural poor, including females.

In most developing countries, female-run enterprises tend to be undercapitalized, and to have poorer access to machinery, fertilizers, information and credit than male-owned enterprises do. Laws, regulations, and customs restrict women's ability to manage property, conduct business, or even travel without their husbands' consent.

Disparities also exist in women's workforce participation. Women are three times as likely as men to be hired informally, and are much more likely to be among the unpaid workers who contribute to the family's business [28].

Such discrepancies impair women's ability to participate in development and to contribute to higher living standards. From a competitiveness perspective, women's disproportionately low participation in the workforce can reduce the pool of applicants and distort the allocation of talent and the productivity of human capital, thereby reducing the average productivity of the labor force [28].

Recently, the increase in the number of female-owned enterprises is much higher than that of their male counterparts in developed countries, according to Niethammer 2013. Furthermore, there is no evidence that women-owned enterprises fail at a faster rate. In this respect, it is important to mention that the percentage of women, who start their businesses motivated by an opportunity, is higher in high-income countries than in low-income/middle-income country groups; meanwhile, in the latter case, women entrepreneurs are more likely to start their businesses just for necessity reasons.[18]

A key impetus of the study at hand is the substantial percentage of female owned SMEs in emerging countries. Although they don't constitute the majority of SMEs owners in developing countries, female-owned small and medium enterprises (SMEs) represent 30 to 37 percent (8 million to 10 million) of all SMEs in emerging markets, according to the IFC in 2013. Several case studies show that ICT can, and actually is, empowering women in developing countries [2]. For example, ICT provides women entrepreneurs with access to worldwide e-business channels, which can be operated 24 hours a day from home in real time [11]. Tele-working, call centers, the software industry and offshore services all call for more IT education and training in all levels of education, which would enhance girls and, later on, women, to become active contributors to the Egyptian economy's growth and development. [2]

1A   1B

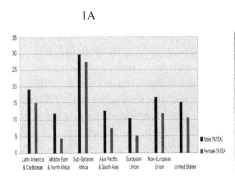   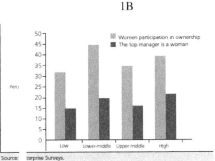

Fig. 1A. Distribution of male/female enterprises according to region

Fig. 1B. Women participation in ownership and top management

Source: GEM Report 2012

In Figure 1A, according to the 2012 Global Entrepreneurship Monitor Report, [7]female entrepreneurship rates are significantly high in Latin America and the Caribbean region, and are relatively low in the Middle East and the North Africa regions. An important trend in this respect is that the number of female entrepreneurs reflects the general trend for entrepreneurship conditions in the specific country. In addition, Latin America and the Caribbean region are efficiency driven economies, while the Middle East/North Africa (MENA) regions contain predominantly factor-driven economies [7]. Figure 1B is taken from to the World Bank's Enterprise Surveys (ES)[29-33]. It depicts that for those females who are engaged in SMEs, female participation in ownership is much higher compared to the females in top management, especially in developing countries compared to other regions in the world.

Policymakers are progressively investigating and developing new ways of empowering women by promoting economic activity and growth among them in developing countries. Furthermore, the increased uptake of access ICTs such as access and use of the Internet and Broadband, having a website for the small businesses, using emails, etc. [29], should be contributing to overall growth and increased productivity level of female-owned enterprises. This is quite important and relevant to empowering females in general and female-owned entrepreneurs *per se*, as those SMEs, that adopt these new techniques of doing business, enjoy a competitive edge compared to the rest of the SMEs that don't incorporate ICT in their production technique. Other ways that ICT can contribute to the growth of the female-owned SMEs include tele-working, e-commerce, and outsourcing[15]. .

On the macro-level, which has been well documented and researched, the vital role of ICT in development allows the developing and emerging countries to leapfrog and excel in their economic and social development (for example[2];[1]; [19]) Thus the question remains whether these technologies can help female-owned SMEs to overcome their disadvantages and become more productive and profitable entrepreneurs.

This research paper embarks on a comparative study based on empirical techniques to investigate the impact that ICT plays on empowering women entrepreneurs in 5 developing/ emerging countries, namely Egypt, Jordan, Morocco, Algeria, and Brazil. The majority of the empirical literature on entrepreneurship discusses solely the hurdles that male entrepreneurs might be confronting when establishing SMEs. [4] . Little empirical evidence has been found on the impact of ICT services such as Internet access and use on the economic performance of female-owned SMEs.

In addition, studies that have investigated the gender dimensions and explored the topic of women entrepreneurs have concentrated on developed countries, as well as the determinants of female entrepreneurs [22].

Thus, the research question of the present study can be summarized by the following questions:

- Does ICT play a role in empowering women entrepreneurs in these Arab countries and Brazil?
- Examining the lessons learned from Brazil's experience, what is the difference in ICT impact on entrepreneurship between the two regions?

It is worth noting that the present research has a twofold value added: it focuses on the research question of empowering female entrepreneurship in Arab countries and Brazil. Further, it adds a new dimension to the existing literature by focusing on the role that ICT is playing to empower these female owned SMEs in the two regions considered.

In the next section, we examine the main characteristics of female entrepreneurship in Brazil and in Arab countries. This is followed by a survey of the pertaining literature. Then, we present a description of the data and methods used to estimate the suggested indicators that measure the performance of female-owned SMEs. Finally, the results are discussed and analyzed. We conclude with a summary of the key findings and suggested policy recommendations.

2   Female Entrepreneurship in Brazil and Selected Arab Countries

In the outset, in the selected Arab countries combined female–owned SMEs constitute approximately only 15% of the total entrepreneurs, while females in developing countries own around 34.3 percent of the small firms [13]. Thus the female–owned SMEs in the selected Arab countries are below the indicated level. However, in Brazil female-owned SMEs are accounting for more than half of the total SMEs owners. Secondly, females *per se* tend to own firms in the services sector and they are smaller in terms of sales, assets, profits, and employees. Finally, literature refers to the fact that entrepreneurs' wage gap between male and female is large [13].

Female Entrepreneurship in Brazil constitutes an important benchmark for the female- owned SMEs in other developing countries such as Arab countries. Most businesses in Latin America and in Brazil are micro- and small enterprises. The motivation for most entrepreneurs in this region is the necessity or finance, rather than the opportunity. Small enterprises in Latin America have an immense impact on job creation, and economic diversification. Middle class firms in Brazil specifically are recognized for being more involved in creating knowledge–based companies (i.e. new communication and software firms as well as internet-related services, and other branches of applied electronics) (World Bank 2013). The present paper draws attention to Brazil based on the following reasons: Brazil is one of the leading emerging countries (BRICS), and it is a leader in entrepreneurship, with an estimated one in eight adults being entrepreneurs. Much of the business that occurs in Brazil is done by single business people either selling their homemade goods or providing their services. Also half of the entrepreneurs in Brazil are women (46%) in 2004, while 4 years ago it was only 29%. New programs have recently been started to help women become even more involved in business. Opportunity-based entrepreneurship and need-based entrepreneurship are close to being equal [7].

One of the major hurdles in starting a new enterprise in Brazil is paperwork. Thus, bureaucracy and gender credit gap in Brazil are considered major hurdles. Despite this, Brazil ranks very high on the Global Entrepreneurship Monitor (GEM) list of countries with the highest entrepreneurial activity (International Entrepreneurship.org).

It is worth mentioning that Brazil ranks high (62) in the Gender Gap Index 2013, where it has succeeded in closing the gender gap in both sub indexes of the health & survival and education attainment.[8]

Female-owned enterprises in Arab countries tend to choose the service business sector [20]. Mostly, the services sector can provide better working conditions than agriculture, and more chances for women to be promoted professionally. The size of their business is relatively small, and they are prone to employ fewer employees. The latter study has empirically proved that the average size of female-owned enterprises in Bahrain and Oman is small. The number of personnel employed by the women entrepreneurs in both Bahrain and Oman ranges between 2 to 60 people [17]. Generally, the main features of the services sector is the low initial requirement of capital and thus low barriers to entry and easier way of starting a business[22]. In general, the services sector can provide better working conditions than agriculture, as well as more chances for women to be promoted professionally [13]. According to the latest GEM Report 2012, in the Middle East and North Africa region, males have demonstrated four times the likelihood of entering a business compared to females. Women entrepreneurs in this region have large families, with an average of five people per household, and they operate primarily as one-woman businesses with no employees.

However, being a women entrepreneur in a developing country entails many hurdles that are worth investigating. These hurdles include limited access to credit, limited pool of human capital, hampering legal environments, and very expensive and complicated ways to have access and use of ICT to promote their businesses. In the latter case, the impact of having access and using ICT for women-owned SMEs has not been well researched in empirical studies, especially in developing countries, such as the above-mentioned Arab countries and Brazil.

Other external factors that form an obstacle to Arab female entrepreneurship include family and community opinions and social norm [2]. It is imperative in the context of the current research to highlight the relationship between women Entrepreneurship and business climate. In the economic aspect, we find that women frequently face gender bias in the socioeconomic environment when it comes to establishing and developing their own enterprises and accessing economic resources. These are not only disadvantages to women, but they also reduce growth potential, productivity, and performance of the economy as a whole. Gender-based inequalities impose significant development costs on societies. Women are active economic participants as business owners, workers, and managers globally. Apparently, there are positive correlations between women's representation on corporate boards and corporate performance, suggesting that women are good for business. Fortune 500 firms with the highest percentage of women corporate officers yielded on average 35.1 % greater return on equity and 34 percent greater return to shareholders than those with the lowest percentage of women corporate officers [26][27].

3   Literature Review

There are only few empirical studies investigating the impact of use and access of ICT in female-owned enterprises in general, not to mention Arab or emerging countries.

Examining this topic across the developing countries, there are a few studies that are worth mentioning. One of the most notable recent papers in this area of research was developed by the World Bank team of researchers, [4] . The paper discusses the performance gaps between male-owned and female-owned enterprises using the ICA surveys but it is confined to four regions, namely Eastern Europe and Central Asia, Latin America, and Sub-Saharan Africa. They report that on average, female- owned enterprises are significantly smaller in terms of overall sales than those of their male-owned enterprises in each region. The concentration of women and men in specific industrial sectors follows very similar patterns in all regions, and there was no strong evidence of credit constraints. Some exceptions were found in the Latin America region. They conduct an empirical analysis using a multiple regression model to estimate their model. They also examine gender gaps in firm growth over a 3-year period, both in terms of employment and sales. In addition, they measure the gender gap and its impact on productive efficiency in terms of the revenue that they generate from given inputs.

An important paper [6] focuses on measuring the effect of ICT on the economic performance of SMEs in three East African countries: Kenya, Tanzania and Uganda. The study uses diffusion of ICT among East African SMEs, which is both industry and country specific, where industries covered in the analysis include food processing, textile and tourism. In their empirical analysis, they tackle three performance indicators: internal rate of return, labor productivity, and domestic and export market expansion. Data obtained from ICA surveys covers East Africa, especially Kenya and Tanzania. The survey that was carried out from November 1999 to May 2000 includes enterprises from three sectors: food processing, textiles and tourism. The entire sample includes 300 enterprises. The distribution is 150 enterprises from Kenya and Tanzania each, distributed equally among the three sectors. In the selection of enterprises, the survey followed a simple random sampling procedure where the sample enterprises are randomly selected from major commercial corridors in the countries. Findings reveal that investment in ICT has a negative impact on labor productivity and a positive impact on general market expansion. However, such investment does not have any significant impact on enterprises' return, nor does it determine enterprises exporter (non-exporter) status[6].

As to the female-owned SMEs performance in developed countries such as in the United Kingdom and the USA, another strand of literature is under consideration. One paper that discusses the gender aspect in firm performance is the one by Robb *et al* 2012[20]. Although in their paper they focus on the US, many important conclusions are revealed regarding the impact of gender on SMEs performance. They come to the conclusion that females in the US are not in a disadvantageous status in terms of necessary skills and financial resources to open their enterprise. Thus, according to this study, there is no empirical evidence that female-owned SMEs underperform male-owned ones because they are smaller or because females in the US prefer to take fewer risks. Gender has no impact on the performance of the firm, and the differences within each gender are much larger than the average differences between the two genders.

Papers that have tackled the ICT impact of female-owned SMEs are scarce. One study by Martin *et al* 2005 [16] , investigated how ICT and the Internet influence the development of women entrepreneurs in the UK using qualitative techniques. This paper's main focus is on female-owned ICT small enterprises. It has found that female-owned SMEs gained advantage by experimenting with new ideas at work and at home. ICT and the Internet empowered female-owned SMEs, where they became more autonomous and organized in their work. New business opportunities were explored with the help of ICT. Outsourcing, for example, were considered an important option to the female-owned ICT firms, where emphasis was put on new ideas and innovation. In general, ICT enabled these female entrepreneurs to become low-cost producers, specializing for differentiation and reaching more markets due to the adoption of the Internet in their businesses, making ICT a key factor in their improved performance and running the business. Examples of new techniques such as tele-working and e-commerce were cited in this research.

Thus we can conclude that there exists a knowledge gap, and an ambiguous analysis, regarding the relationship between female owned SMEs performance, and the role that ICT is playing in empowering them. ICA surveys had already been used by several studies to investigate the performance of SMEs across various regions. However, a new perspective, that addresses gender dimension and how ICT can be a tool or a vehicle used to empower female-owned SMEs, has not yet been rigorously researched or examined using the ICA surveys.

4   Research Methodology

Data:

This research uses the World Bank's Enterprise survey or the Investment Climate Assessment Survey (ICA) as the source for data,   for the four Arab and emerging countries, namely Egypt (ICA 2008), Jordan (ICA 2006), Morocco (ICA 2007), Algeria (ICA 2007), and Brazil (ICA 2009). The ICA database provides comparable enterprise level data because it uses similar sampling techniques (ES Global sampling methodology). The definition of SMEs is uniform across all sectors in the Investment Climate Assessment (ICA) Survey in all sectors, in which small firms have 5-19 employees; medium–size firms are ones with 20-99 and firms with 100 or more employees are classified as large. One of the points of strength of the ICA surveys includes identifying the gender of the principal owner of the enterprise, the size of the firm and its age thus far. Another outstanding fact about these investment surveys includes their completion on a regional level. Thus we can compare countries or a group of countries of two different regions such as the Arab countries and Brazil. The Enterprise Surveys cover small, medium, and large companies. The surveys are conducted to a representative sample of firms in the non-agricultural formal private economy.   (http://www.enterprisesurveys.org)[29-33]. In this study the data for four Arab countries were pooled, because of the small data set available for each individual country, so the pooling served the purpose of increasing the sample size.

Services sector in the ICA survey includes the following industries (commercials, information technology, construction and building, tourism services and transportation, restaurants and others). The Brazil survey includes both the manufacturing and the services sector. This is due to the fact that ICT is used by both manufacturing and services sectors in Brazil, which is different from the case in Arab countries. The manufacturing sector in Brazil includes food, textiles, garments, shoes and leather, chemicals, machinery & equipment, auto parts, furniture and other manufacturing; and the services sector comprises construction, wholesale, retail, hotels and restaurants, transportation and others.

The Enterprise Surveys provide indicators that describe several dimensions of technology use and innovation. ICT indicators shed light on the use of information and communication technologies (ICT) in business transactions. ICT, such as the Internet, is considered an important tool for all firms, especially SMEs. ICT tools should empower enterprises, especially SMEs, so that they can arrive at national and international markets at lower costs [25] The questions that are extracted from the individual level survey will be used for this study to construct the ICT index. (see Table 2)

The performance of the enterprise is measured using two different indicators, namely the internal rate of return and the labor productivity. Why is Internet such an essential ingredient for a successful SMEs, for example in the Egyptian market? SMEs in general report sizable benefits from having online presence. These include the following advantages: improved returns from their advertising campaign; increased productivity; greater ease in recruiting; enhanced ability to provide a large range of products to customers. The biggest benefit reported by SMEs that are active online is their ability to expand their geographic footprint (especially into other regions in Egypt). However , the biggest challenge is lack of awareness [5].

5   Data Analysis

Preliminary analysis of the data of the four Arab countries: Jordan, Egypt, Morocco and Algeria, reveals the dominance of male entrepreneurs in the selected Arab countries under study, compared to Brazil. In the latter case, we observe that the female-owned SMEs outweigh the male-owned SMEs in number. Analyzing the breakdown of the female-owned SMEs on a country level; we become aware that they are the highest in Algeria, followed by Morocco, Jordan and finally Egypt. This is an unexpected result, since Egypt is the most populous country of the four Arab countries investigated.

Table 1. The distribution of SMEs owners in Selected Arab Countries and Brazil, based on Gender

Country	Male	Female
Algeria	85%	15%
Jordan	85%	15%
Morocco	88%	12%
Egypt	72%	28%
Brazil	44%	56%

Source: ICA Survey

On the other hand, in Brazil we find that female-owned SMEs are much more (614 SMEs) compared to male–owned ones (486 SMEs). We also notice that the major type of enterprises in the selected countries is SMEs.

Moving to the ICT indicators, Jordan is ranked the first in terms of the usage of high speed internet access, usage of Internet for making purchases, usage of Internet for doing research on new products, and usage of the web to communicate with customers. However, when it comes to using emails to communicate with clients, Morocco ranks number one (Table 2). Brazil, on the other hand, shows different results. The number of females who are connected to the the internet in Brazil are higher compared to the males. However, in the rest of the ICT indicators, the males are leading the access and use of the Internet.

As to Algeria & Morocco: It stands out that "the establishment uses E-mail to communicate with clients or suppliers"; this represents approximately 46 % and 57%,

Table 2. The ICT indicators in ICA surveys

Country	The establishment uses website to communicate with clients or suppliers	The establishment uses E-mail to communicate with clients or suppliers	The establishment has high speed internet connection with its premises	The establishment uses internet to make purchases	The establishment uses internet to deliver services	The establishment uses internet to do research and develop ideas on products and services
Algeria	24%	46%	12%	5%	5%	8%
Jordan	20%	26%	20%	15%	9%	9%
Morocco	25%	57%	6%	4%	4%	4%
Egypt	21%	21%	13%	15%	15%	16%
Brazil	33%	50%	5%	4%	4%	4%

Source: ICA Surveys

respectively, of the overall percentage representing the usage of ICT in SMEs measured by the ICA survey. However, in Egypt and Jordan, the share of the various ICT services and products in the idea of including ICT in the SMEs is more evenly distributed among the various services. In Brazil, we also find out the fact that "the establishment uses E-mail to communicate with clients or suppliers" is quite popular among female-owned SMEs. In Brazil, data reveal that the number of females who are connected to broadband are higher compared to the males. However, in the rest of the ICT indicators, the males are leading the access and use of the Internet.

In this respect, it is worth noting that these findings are consistent with the general performance of the ICT sector in the respective Arab countries. According to the ITU data, in 2008 and 2009 Brazil had a surge in Internet penetration rate, as in the one depicted by ITU data (Figure 2).

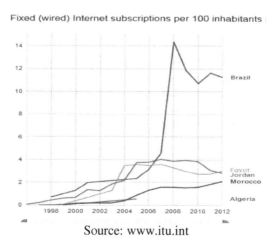

Source: www.itu.int

Fig. 2. Fixed Internet subscriptions per 100 inhabitants in selected Arab countries and Brazil

The Empirical Model

The present paper is based on the same research methodology applied by Chowdhury *et al* 2003 in "Use of ICTs and the Economic Performance of SMEs in East Africa". The new dimension, added to the previously mentioned study, is conducting a comparative study between two different countries, where lessons learned can be drawn from Brazil's experience. The second additional dimension in this study focuses on women entrepreneurs in both countries, thus paying attention to the gender dimension and women empowerment. An ICT index is constructed using the questions pertaining to the following ICT and R&D services, given that R&D uses ICT quite intensively using the Principle Component Analysis Technique, as discussed previously. We will measure the impact of using and spending on ICT for female owned firms on the following two economic performance indicators for the countries under study:

A: The internal rate of return for female-owned SMEs in Arab countries and Brazil

B: The labor intensity and labor productivity or female-owned SMEs in Arab countries and Brazil

The Internal rate of return is our dependent variable and it will be measured using the following formula: Revenue minus Variable costs divided by aggregate capital stock. Then, to measure the impact of ICT on female-owned enterprises on the IRR, the following equation will be estimated using OLS:

$$IRR = \beta_0 + \beta_1 \ln(K/Y) + \beta_2 \ln(ICT/K) + \beta_3 \ln(EQ/K) + \beta_4 \text{Female Owned SMEs} + \beta_5 ICT\text{index} + \varepsilon \quad (1)$$

The first indicator, namely the Internal Rate of Return (IRR), is regressed on aggregate capital intensity (K/Y) or capital deepening, the ratio of ICT capital to total capital and the ratio of non-ICT capital to total capital (ICT/K), which is the method to model capital by breaking down capital into ICT and non ICT assets (EQ / K) [5], in addition, we include other control variables such as female-owned SMEs and ICT index in Arab countries. In Brazil's regression model, we control for using Website to communicate with clients and suppliers variables as a proxy for ICT variables discussed previously, since there are a huge number of missing values in the rest of the ICT variables in Brazil's ICA report. The method of estimation is OLS and the regression is a level–log Regression.

The second performance indicator for measuring the impact of ICT in female-owned SMEs is the Labor intensity variable. In order to measure the impact of ICTs on labor intensity in female-owned enterprises, a similar relationship can be derived. Applying a Cobb-Douglas production function and assuming constant returns to scale, the following equation will be estimated:

$$\ln(L/Y) = \alpha_1 + \alpha_2 \ln(K/Y) + \alpha_3 \ln(ICT/K) + \alpha_4 \ln(EQ/K) + \\ \alpha_5 \text{Female Owned SMEs} + \alpha_6 ICT \text{ Index} + \varepsilon \quad (2)$$

where the independent variable is (L/Y), which measures labor intensity. Method of estimation is OLS, where the regression is a log-log Regression. In the selected Arab countries, we control for the ICT index, while in the case of Brazil we control for one of the ICT variables, namely the establishment uses Website to communicate with clients or suppliers, for the same reason stated previously.

Prior to running the regression analysis, a correlation matrix of the variables was prepared. Table 1 in the appendix shows the means and the standard deviations, and Tables 2 and 3 show the correlations of the variables used in subsequent regression for SMEs in the selected Arab countries and Brazil. The pair wise correlations were not large enough to warrant concern about possible multicollinearity problems. In order to estimate the previous multiple regression models using ordinary least square (OLS), we verified the five assumptions of OLS, namely; linearity, constant variance (Homoscadasticity), Multicollinearity, weak exogeneity, and independence of errors, then we pooled the data and performed the regressions. Pooled cross sections are obtained by collecting random samples from large population independently of each other at different points in time, assuming that they have homogenous slopes. Model 1 captures the impact of female-owned SMEs and ICT among other control variables on the profitability of the SMEs i.e. the internal rate of return in the selected Arab countries (Egypt, Jordan, Morocco and Algeria). The model is overall significant with probability of F statistics equal to zero. For the Arab countries, all predictors in the model 1A have a significant economic and statistical impact on IRR except for the Non-ICT capital, which is statistically not significant. We notice that the female-owned SMEs have a negative association with the internal rate of return or simply their profits are lower compared to male - owned SMEs.

In model 1B, the ICT index is controlled for in the regression in addition to the female–owned SMEs explanatory variable. We notice that the coefficients for the ICT index and the female–owned SMEs are statistically insignificant.

Model 2 captures the impact of female-owned SMEs on labor intensity. There were log transformations to most of the variables in addition to the dependent variable as per the literature (Berndt et al 1992). This is a log–log regression where the estimated parameters are elasticities.

Results reveal that the female-owned SMEs have a negative association with the labor intensity. However, this association is not statistically significant. The ICT index is statistically significant, and has a positive impact on the labor intensity in the selected Arab countries.

As for Brazil, we find in Model (2A), where the dependent variable is the Internal Rate of Return (IRR), and we controlled for female–owned SMEs–and the case of having a web site, we find that there is a negative association, not statistically significant, between the IRR and the female- owned SMEs, and there is a positive association, not statically significant, association between using a website by the SMEs and their IRR.

The labor intensity model reveals that all the controlled variables are statistically significant, except for the female–owned SMEs. However, there exists a negative, statistically significant, association between the labor intensity and the case of having a website to communicate with clients or suppliers.

6   Analysis of the Results

In the selected Arab countries, the increase in female-owned SMEs is associated with a decrease in Internal Rate of Return. However, when we control for ICT in terms of

ICT index constructed using the Principal component analysis technique (PCA), the female-owned SMEs becomes statistically insignificant; the same is true about the ICT index. This implies that IRR is negatively associated with the female owners of the SME, and there is a no- association between IRR and the access and use of ICT. In Brazil, however, neither gender nor ICT played any role in the profitability of SMEs.

However, as for the other measure for economic performance, namely the labor intensity, the findings reveal that in the selected Arab countries and in Brazil, the ICT index has a positive, statically significant, association with labor-intensity; meanwhile in Brazil the usage of Website has a negative, statistically significant, association with the labor-intensity.

The female ownership of SMEs seemed to impact only the internal Rate of Return in the selected Arab countries and it has a negative association. On the other hand, we can conclude that gender plays no role on the economic performance of the SMEs in Brazil.

In order to understand the impact that ICT has on the labor intensity as a measure for economic performance of SMEs, it is worth mentioning that the impact of ICT on labor intensity can be positive or negative. As the productivity of a worker, who works in SMEs, increases, fewer workers are needed to produce one unit of output. If the firms do not change the amount of goods they produce, productivity leads to losses in jobs, which is a negative effect of productivity growth and thus on labor intensity. However, if we conclude that there is a positive impact and productivity increase, this will be due to the fact that an increase in productivity reduces the cost of production per unit. Thus, the price of goods decreases and consumers are able to buy more goods, a case which also increases the demand for these goods. In addition, firms become more competitive in prices at the international level, a case which could result in increased exports and global demands for their goods. As a result, firms can sell and produce more, and therefore may end up expanding and hiring more workers([13]. However, in the long-term, job growth and productivity are positively correlated. Investing in training can help raise firms' profits and labor market outcomes due to positive productivity returns from worker education, especially in the long run. With respect to the control variables, controlling for the enterprise's age did not turn out to be significant in any case, and the results of the analysis were similar to those showed in Table 6. Thus, it was decided to not include it in the models of regression.

7   Conclusion and Policy Recommendations

This empirical study reveals that female-owned SMEs have a negative and significant association on the internal rate of return in Arab countries; meanwhile gender plays no role in SMEs performance in Brazil. ICT shows a positive association with IRR in the selected Arab countries, while negative association is shown in Brazil. Thus, we can conclude that in Brazil ICT replaces, or substitutes, labor and this results in reducing labor intensity; comparatively, in the selected Arab countries, ICT leads to an increase in labor intensity. In addition, we can conclude that due to the various and numerous constraints that women face in Arab countries, such as limited access to commercial credit, lack of required ICT skills, conservative social norms, among other things, female-owned SMEs tend to be less profitable and less productive.

One of the main hurdles facing the female-owned SMEs is the skill mismatch for the owner as well the employees in the enterprise. It is a serious problem that threatens the potential of SMEs, especially in developing countries. It can be ascribed to the deficiency in the education sector and the lagging behind in important advances in technology.

Furthermore Entrepreneurs fall into the trap of the weakness in recognizing the benefits from upgrading and updating their skills according to recent technological advances, such as new technologies in Internet access and mobile communication and usage of smart phones and social media to promote their businesses. Thus, female-owned SMEs could help themselves if they invest more in human capital, updating their skills and using new technologies to improve their business performance.

It is worth noting that in Brazil, policy makers made concrete milestones in empowering women in the Brazilian society, in terms of giving mothers financial support conditional on them being responsible for their children's health and education progress (Bolsa Familia program). This resulted in Brazilian women contributing more to the economic development of their country and having more chances to succeed in their businesses. Only through the existence of the political will, women can be empowered to play a major role in their countries' development process.

Policy makers in Arab countries face the critical responsibility of addressing some policy relevant issues, such as incorporating the women's entrepreneurial dimension in the formation of all SMEs-related policies. This can be done by ensuring that the impact on women's entrepreneurship is taken into account in the design stage of SMEs-related policies. It is thus recommended that ICT policies drafted by the designated governmental entities should support the women entrepreneurship in the Arab countries. The awareness of gender issues is important when considering strategies to improve the business environment and promote private-sector development. Mainstreaming gender and creating greater economic opportunities for women have compelling economic reasons, to the extent that inequalities impose development costs on the society. Thus we can conclude that socio- cultural context as well as institutional context play a major role in the success of female entrepreneurs.

Thus, suggested steps to enhance women's participation in economic activities in the Arab countries include the following:

- continuing to improve the business environment and giving females in Arab countries more responsibilities in the society as well as financial support (Drawing from the Brazilian experience);
- addressing norms, traditions and legal discrimination; For example legal and political empowerment of female in terms of giving them more authority to make decisions regarding their children's education and health aspects.
- encouraging women to join the labor force;

- improving employment conditions for female workers;
- increasing women's access to finance, including access to Microcredit; which entails an institutional reform in banking system to encourage female-owned SMEs to obtain lines of credit.
- enhancing vocational training for girls and training for female workers; especially given the fact of low female unemployment rates among vocational school graduates.
- eliminating digital illiteracy among women, and raising awareness of the merits of incorporating ICTs in the management of SMEs;
- giving incentives to women entrepreneurs who already adopt new technologies in their businesses, such as tax reductions or cash loans to upgrade their businesses;
- providing role models

Limitations of this study include the treatment of all Arab countries as homogeneous group of countries as well as not controlling for regional or inter-state differences across Brazil i.e. of how female empowerment varies according to the region in Brazil. In addition, data limitations related to absence of answers to some of the ICT questions in ICA survey of Brazil of 2009.

References

1. Badran, M.F., Hosein, H.F.: Promoting Economic Growth by Broadband Development in Emerging countries: An Empirical Study, Telecom World (TW), ITU (2011), http://www.ieeexplore.ieee.org (retrieved)
2. Badran, M.F.: Is ICT Empowering Women in Egypt? An Empirical Study. In: Proceedings of the Research Voices from Africa Workshop, IFIP WG 9.4, Markerere University, Uganda (2010)
3. Badran, M.F.: What Determines Broadband Uptake in Emerging Countries? An empirical study. In: IFIP 9.4. Conference, Assessing the Contribution of Computers in Achieving Development Goals, Dubai, UAE (2009)
4. Bardasi, E., Sabarwal, S., Terrell, K.: How do female entrepreneurs perform? Evidence from three developing regions. Small Business Economics, Springer Science and Business Media 37, 417–441 (2011)
5. Berndt, E.R., Morrison, C.J.: High-tech Capital Formation and Economic performance in US Manufacturing Industries: An Exploratory Analysis. Journal of Econometrics 65, 9–43 (1995)
6. Boston Consulting Group, Egypt at a Crossroads How the Internet Is Transforming Egypt's Economy Report (November 2012)
7. Chowdhury, S.K., Wolf, S.: Use of ICT and the Economic Performance of SMEs in East Africa United Nations University, World Institute for Development Economics Research (WIDER), Discussion Paper No. 2003/06 (2003)
8. Global Entrepreneurship Monitor Report, Global Report (2012)
9. Gender Gap Report, World Economic Forum (2013)
10. El-Hamidi, F.: How Do Women Entrepreneurs Perform? Empirical Evidence from Egypt. Alma Laurea Working Papers 23: Bologna (2011)
11. Heckman, J.: Sample selection bias as a specification error. Econometrica 47(1), 153–161 (1979)

12. Hilbert, M.: Digital gender divide or technologically-empowered women in developing countries? A typical case of lies, damned lies, and statistics. Women's Studies International Forum 34(6), 479–489 (2011)

13. I.F.C. Issue Brief / Women and Business, http://www1.ifc.org/wps/wcm/connect/bb0b20004d0481febbebfff81ee631cc/IFC-issue-Brief_AM12_Women-and-Business.pdf?MOD=AJPERES

14. IFC Jobs Study, ASSESSING PRIVATE SECTOR CONTRIBUTIONS TO JOB CREATION AND POVERTY REDUCTION. IFC Report: Washington DC, USA (2013), http://www.internationalentrepreneurship.com

15. Klapper, L., Parker, S.: Gender and Business Environment for New Firm Creation The World Research Observer. Oxford University Press, USA (2010)

16. Maier, S., Naier-Reichert, U.: Empowering Women Through ICT-Based Business Initiatives: An Overview of Best Practices. E-Commerce/E-Retailing Projects 4(2), 43–60 (2007)

17. Martin, L.M., Wright, L.T.: No gender in cyberspace? Empowering entrepreneurship and innovation in female–run ICT small firms. International Journal of Entrepreneurial Behavior & Research 11(2), 162–178 (2005)

18. Nasr, S. (ed.): Egyptian Women Workers and Entrepreneurs. Maximising Opportunities in the Economic Sphere. World Bank, Washington DC (2010)

19. Niethammer, C.: Women Entrepreneurship and the opportunity to promote development and Business. Brookings Blum Roundtable (2013)

20. Qiang, C.: Broadband Infrastructure Investment in Stimulus Packages: Relevance for Developing Countries. World Bank, Washington D.C (2009)

21. Robb, A.M., Watson, J.: "Gender differences in firm performance: Evidence from new ventures in the United States. Journal of Business Venturing 27, 544–558 (2012)

22. Teltscher, S.G., ICT and Development: Electronic Commerce and Development Report, UNCTAD: Geneva, Switzerland. 2002

23. Verheul, I., Van Stel, A., Thurik, R.: Explaining female and male entrepreneurship at the country level. Entrepreneurship & Regional Development 18, 151–183 (2006)

24. World Bank. Engendering development through gender equality in rights, resources, and voice. World Bank Publication (2001)

25. Lora, E., Castellani, F.: World Bank & Inter-American Development Bank: Entrepreneurship in Latin America, A Step Up the Social Ladder (2013)

26. World Bank, Mary Hallward-Driemeier. Enterprising Women, Expanding Economic Opportunities in Africa. African Development Forum (2013)

27. World Bank Report. Capabilities, opportunities and participation: gender equality and development in the Middle East and North Africa region. World Bank Publication (2011)

28. World Development Report: Gender Equality and Development (2012)

29. World Bank : Investment Climate Assessment Survey (ICA) for Egypt, Washington DC (2008)

30. World Bank : Investment Climate Assessment Survey (ICA) for Algeria, Washington DC (2007)

31. World Bank : Investment Climate Assessment Survey (ICA) for Jordan, Washington DC (2006)

32. World Bank : Investment Climate Assessment Survey (ICA) for Morocco, Washington DC (2007)

33. World Bank : Investment Climate Assessment Survey (ICA) for Brazil, Washington DC (2009)

Annex 1

Table 3. Correlation Matrix

	IRR	Ln L/ Y	Ln K/Y	Ln ICT /K	Ln Non ICT/K	FSMES	ICT Index
IRR	1.0000						
Ln L/Y	0.1180	1.0000					
Ln K/Y	0.4012	0.5473	1.0000				
Ln ICT/K	-0.1935	0.1630	-0.5031	1.0000			
LN Non ICT /K	0.1698	-0.3329	0.0761	-0.6646	1.0000		
FSMEs	0.0653	-0.0133	0.0479	0.0133	0.0647	1.0000	
ICT Index	0.0688	0.1892	0.1077	0.0643	-0.1084	0.2351	1

Table 4. Descriptive Statistics

Variable	Obs	Mean	Std.Dev.	Min	Max
IRR	929	-40.87816	1251.523	-101540.99	32847.36
Ln K/Y	1085	-3.027191	2.309253	-9.400566	6.327937
Ln ICT/K	976	-2.008448	2.123184	-9.723474	0
Ln Non ICT/K	841	-0.2260372	0.474148	-5.433059	0
Ln L/Y	1757	-12.12837	0.474148	-25.04858	2.690088
FSMES	1754	0.1510832	0.3582321	0	1

Table 5. Dependent variable Internal Rate of Return: (IRR) in female-owned SMEs in Selected Arab Countries

Model (1A)
Level –Log Regression

Explanatory Variables	Parameter	t-statistics	P- value
Constant	152.2638***	4.29	0.000
Ln K/y	34.66043***	5.69	0.000
lnICT/K	20.6042***	2.90	0.004
Ln Non ICT /K	25.00185	1.35	0.179
FSMES	-69.28906***	-3.08	0.002

Number of observations: 472
Adj-Rsquared= 10%
Prob>F=0.000
*$p < 0.1$.
**$p < 0.05$.
***$p < 0.01$

Table 6.

Model (1B): Dependent Variable: Internal Rate of Return (IRR) in female–owned SMEs in Selected Arab Countries controlling for ICT Index
Level-Log Regression

Explanatory Variables	Parameter	t-statistics	P- value
Constant	113.8147***	2.77	0.007
Ln K/y	27.57132***	4.47	0.000
Ln ICT/K	14.8127	1.53	0.129
Ln Non ICT /K	0.1142882***	2.32	0.022
FSMES	2.356242	0.10	0.922
ICT Index	1.73022	0.19	0.849

Number of observations: 111
Adj-R squared: 17%
P>F-stat: 0.000
$*p < 0.1$.
$**p < 0.05$.
$***p < 0.01$

Table 7.

Model 2: Dependent in (Labor intensity)
Log-Log Regression

Explanatory Variables	Parameter	t-statistics	P- value
Constant	-2.887595***	-3.24	0.000
Ln K/y	1.165607***	9.54	0.000
lnICT/K	0.2102365***	5.05	0.000
Ln Non ICT /K	0.2102365	0.29	0.769
FSMES	-0.6092088	-1.50	0.137
ICT Index	0.2741077*	1.70	0.093

Number of observations: 107
Adj-Rsquared= 52%
Prob>F=0.0000
$*p < 0.1$.
$**p < 0.05$.
$***p < 0.01$.

Brazil:

Table 8. Model (2A): Dependent Variable: Internal Rate of Return (IRR) in female–owned SMEs in Selected Brazil controlling for Web

Explanatory Variables	Parameter	t-statistics	P- value
Constant	3224.431	3.52	0.000
Ln K/y	1163.33***	7.27	0.000
lnICT/K	-45.92594	-0.16	0.874
Ln Non ICT /K	847.5376***	3.55	0.000
FSMES	-730.9831	-1.37	0.172
Web	35.07294	0.06	0.950

Number of observations: 467
Adjusted R-squared: 10%
P>F-stat.=0.0000
*p < 0.1.
**p < 0.05.
***p < 0.01Table 9:

Model (2B): Dependent Variable: Labor Intensity (ln L/Y) and controlling for female–owned SMEs and having a website

Explanatory Variables	Parameter	t-statistics	P- value
Constant	-9.627593***	-41.42	0.0000
Ln K/Y	0.5930852***	14.43	0.0000
lnICT/K	0.1551133**	2.06	0.0400
Ln Non ICT /K	0.2281846***	3.80	0.0000
FSMES	0.2039671	1.50	0.135
Web	-0.6261691***	-4.32	0.0000

Number of observations: 541
Adjusted R-squared: 35%
P>F-stat.=0.0000
*p < 0.1.
**p < 0.05.
***p < 0.01

Celebgate: Two Methodological Approaches to the 2014 Celebrity Photo Hacks

Rebecca Fallon[✉]

Oxford Internet Institute, Oxford, UK
rebecca.fallon@oii.ox.ac.uk

Abstract. On August 31st, 2014, nearly 500 sensitive images captured from the mobile phones of various celebrities were released onto 4chan.com. With alarming alacrity, these stolen personal photographs made their way to slightly more mainstream content sites, including Reddit, Tumblr and Twitter. Internet users and media respondents have termed the phenomenon "Celebgate" or, more popularly and vulgarly, "The Fappening" (a portmanteau between 'happening' and 'fap'—slang for masturbation). The leak raised numerous questions about privacy rights online, iCloud security, and the responsibilities of host sites. This paper will examine two research designs for the study of this viral content; the first considers the leak structurally and examines the relationship between search terms and centralization of information, the second considers the leak morally and questions variation in the relationship between the purported morality of a website and those of its users. This is primarily an exploratory methodological paper; and the sensitivity of the topic brings its own limitations to the types of data a researcher can ethically and practically capture.

Keywords: Search · Privacy · Viral

1   A Quantitative Approach

Viral content generally lends itself well to big data studies because it leaves a long trail. Viral video phenomena like PSY's "Gangnam Style" or the meme "Harlem Shake" that spread across Twitter can be mapped across the platform and analyzed in terms of geographic, temporal, and nodal diffusion.[1] Because this content is all unquestionably public, the persistent metadata it generates (i.e. time, location, originator of post) can be easily scraped from social platforms. However, in the case of our viral celebrity photo outbreak, the sensitivity of the content demanded its censorship on some of the most prominent sites to host it. While this censorship of stolen private property may be morally commendable, it does present some challenges to what information can be accessed about the spread of the phenomenon.

---

[1] D'Orazio, Francesco. (2013, May 8). "How Stuff Spreads #1: Gangnam Style Vs. Harlem Shake—Full Study and Data Visualization." *FACE* blog. Online: http://www.facegroup.com/blog/how-stuffspreads-1-gangnam-style-vs-harlem-shake.html

© Springer International Publishing Switzerland 2015
T. Tiropanis et al. (Eds.): INSCI 2015, LNCS 9089, pp. 49–60, 2015.
DOI: 10.1007/978-3-319-18609-2_4

Different platforms responded with different stances towards removing the leaked content. Most of the requests for removal cited the Digital Millennium Copyright Act rather than privacy rights, as it is some of the toughest legislature available for forced content removal. Following suit, sites had to adopt the language of property rather than privacy in justifying their policies. Reddit, which quickly became a principle hub for hosting the stolen images, shut down the dominant forum (or subreddit) centralizing them about a week after their initial upload. They stated publically on their blog that, "In accordance with our legal obligations, we expeditiously removed content hosted on our servers as soon as we received DMCA requests from the lawful owners of that content, and in cases where the images were not hosted on our servers, we promptly directed them to the hosts of those services."[2] In removing the pivotal nucleus for these images, Reddit decentralized ways in which Internet users could find them. Similarly, for the first time in its eleven-year history, 4chan also acknowledged compliance with the DMCA, though the effectiveness of their new self-policing system remains to be seen.

Google was faced with similar legal pressure to remove links to the stolen content. Yet while they publically stated that "We're removing these photos for community guidelines and policy violations (e.g. nudity and privacy violation) on YouTube, Blogger and Google+. For search we have historically taken a different approach as we reflect what's online—but we remove these images when we receive valid copyright (DMCA) notices."[3] However, in practice, Google has allowed many links to stolen content to remain findable in their results. For instance, Justin Verlander, the boyfriend of hacking victim Kate Upton, filed a copyright claim with the request to remove 444 URLs hosting Upton's stolen photos. In a transparency report, Google states that no action was taken in 41% of these cases; 41% of the images requested for removal still appear in Google search results.[4] While they did remove the majority of links to the most central sites, they allowed many of the more peripheral domains to remain. For instance, though Reddit removed The Fappening, Google could still reveal links to Upton's photos on subreddits like /r/Celebs or /r/ledzepplin.

Bearing in mind this discrepancy in censorship as well as the knowledge that Google is typically the first port of call for enquiries online, both for specific host websites and detached content, we arrive at an interesting research question. *How does centralization and decentralization of content create a shift in search terms?* Specifically, when Reddit shut down /r/TheFappening, how did individuals adapt their Google search terms to locate the decentralized content? This research question would require a trend analysis of different search terms given a timeline of major host censorship.

---

[2]   "Every Man Is Responsible For His Own Soul" (2014, 6 September). *Blog.Reddit.* Online:
      http://www.redditblog.com/2014/09/every-man-is-responsible-for-his-own.html
[3]   Stern, Marlow. (2014, October 5). " 'The Fappening 4': More Celeb Nudes Leak Online
      Including the First Man, Seems to Be Dying Down." *The Daily Beast.* Online:
      http://www.thedailybeast.com/articles/2014/10/05/the-fappening-4-more-celeb-nudes-leak-
      onlineincluding-the-first-man-seem-to-be-dying-down.html
[4]   Google. (2014, December 1). "Transparency Report; Request ID: 1445734" *Google.* Online:
      http://www.google.com/transparencyreport/removals/copyright/requests/1445734/

An analysis of the search terms is interesting, because it provides insight into not only how a viral phenomenon diversifies or concentrates, but also how individuals must conceptualize the phenomenon in order to access it. Previous studies on search patterns surrounding events are numerous, from work on seasonal queries[5] to tracking the spread of diseases.[6] It would be quite another thing to track the *actual* content as it spread through different domains. This sort of mapping might be easier to accomplish with texts, which tend to be far more searchable than images. Additionally, mapping the spread of content within a specific platform (like Twitter) is far easier than mapping content spread across the web at large—it is nearly impossible to determine exactly where someone picked up a viral image. Images are not only picked up from host sites, but can be sent through email, USB drives, mobile devices or messaging clients. Also, as many host sites rigorously remove this sensitive content, and many of those sites are themselves removed for one reason or another, the trail these photos leave are likely full of (literal) missing links. For this reason, choosing to look not at *where* people ultimately locate content but *how* they try to get there may be a stronger and more productive option for quantitative research on this topic.

Google itself provides tools for examining trends on the site, and allows for an indepth comparison of interest in topics over time and across geographical regions. Using these, paired with tools like the Internet Archive which map popularity of certain webpages over time, could provide interesting visualization and statistics for changes in searches after centralized censorship. They would also offer a nice mechanism for comparison between different kinds of search terms, ie. host platforms vs. decentralized content. Hypothetically, if someone knew he could no longer access certain content on Reddit, he might instead look for the content itself in a more detached form, like 'Jennifer Lawrence Photos.' This methodology could also reveal the extent to which interested Internet users are committed to finding content at all if it is not made easily available on a centralized platform. Would people still continue to search for the photos if they were not aggregated on an established, centralized source? What is the threshold of effort that people are willing to put in to search for content that isn't made as easily findable? This type of trend analysis could give some interesting results towards answering these questions.

One problem with analysis of search terms about such a taboo viral phenomenon is the inability to distinguish between those looking for the content itself and those looking to learn about the leaks without actually consuming the photos. If a person searches "Celebgate" in Google, he could equally be trying to find the *Daily Mail* article on the matter as the actual pictures themselves. Perhaps one could infer based on connotations of the terms used, but those assumptions are risky ones. Thus, a weakness of this approach is the inability to separate the phenomenon itself from the meta-phenomenon of its public interest and media response.

---

[5] Shokouhi, Milad. (2011) "Seasonal Queries by Time-Series Analysis" *SIGIR*. Available Online: http://research.microsoft.com/pubs/150746/sigir0196-shokouhi.pdf

[6] Valdivia and Monge-Corella (2010) "Diseases Tracked by Using Google Trends, Spain" *Emerg Infect Dis.* Jan 2010; 16(1). Pg 168.

Moreover, it is difficult to gauge the actual *effectiveness* of a given search term for finding the viral content, though one may be able to guess given the longevity of its popularity. While more people may search "Jennifer Lawrence Photos," it could be that "Jennifer Lawrence pics" yields a greater number of relevant and direct results. This is particularly salient given the intensity with which these search terms spiked and peaked. "The Fappening" (for example) may not have yielded the best results immediately, but quickly became a term around which results could aggregate. In this sense, our method is also limited in its ability to reveal findability of content under a given search term, and one must be cautious in specifying that its results reflect what is *searched for*, not necessarily what is *found*.

An additional weakness could be the assumption that Google is the first port of call for the particular demographic of Internet users interested in these photos. Yes, Vaidhyanathan hardly exaggerates when he writes, "Google has permeated our culture…[it] is on the verge of becoming indistinguishable from the Web itself."[7] However, it is entirely possible that this particular subset of Internet users—many of whom may be more or less peripherally connected to hacking communities already— could use alternate means of finding their content, using browsers like Tor and its accompanying TorSearch. Especially in light of recent public awareness of search engine surveillance and collective suspicion about data collection, Internet users (and especially ones looking for illicit content) might be more wary than ever about what they type into Google. Therefore, Google could provide a very skewed sample of those interested in search terms.

Despite certain limitations on the forms of trend analysis proposed here, such methods can still elucidate important findings about what people search for in response to decentralization of viral content. It may also shed light on the function of naming a phenomenon (i.e. Celebgate or The Fappening) and the relationship between appellation and online findability. More philosophically, we can speculate at how naming something reframes the way that we think about it.

1.1 Possible Findings

Preliminary investigation reveals interesting findings related to I) queries for platforms vs. queries for content and II) queries about aggregated content vs. queries about more specific photographs. The first chart looks at two 'content' terms, "fappening" and "Jennifer Lawrence photos" and searches two of the most prominent platforms hosting the stolen content, 4chan and Reddit.

While Reddit, 4chan, and "Jennifer Lawrence photos" spiked immediately during the first outbreak of photos, it took a bit of a lag for "fappening" to peak. This could have something to do with the newness of the term as it related to the phenomenon. However, during the second outbreak of photos on September 21st, the established term "fappening" rose at the same rate as 4chan and Reddit. This could be an indicator that once the leak was classified with a name, the images began to be

---

[7] Vaidhyanathan, Siva. (2011). *The Googlization of Everything (And Why We Should Worry)*. University of California Press: Berkeley: 2-3.

searched collectively rather than by specific victims of the theft—we can see searches for Jennifer Lawrence's photos drop below searches for the movement as a whole. Moreover, while 4chan was the first port of call during the first leak, Reddit was more highly searched during the second wave. There could be many reasons for this, one of which could be the more 'mainstreaming' of the phenomenon; while 4chan prides itself on its obscurity, Reddit proclaims itself "The Front Page of the Internet."

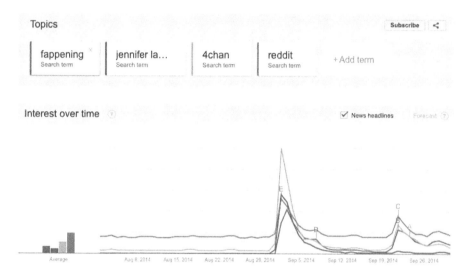

Fig. 1. Searches for "fappening", "Jennifer Lawrence photos," "4chan" and "reddit" between August 1 and September 31, 2014.

The next cursory graph looks at "fappening" as a term, and then search terms for four different celebrities affected by the photo leak.

Searches for both "Jennifer Lawrence" and "Kim Kardashian" are notably higher than searches for "fappening," and specific peaks for each celebrity correspond with the release of their own private photos. Looking at this graph, it would appear that while certain individuals generated interest on their own, "The Fappening" as a phenomenon gained notable interest as well. These findings could have interesting implications for how victims of the cyber-thefts are viewed; as individuals or part of a class. Additionally, the peaks for queries on each celebrity are surprisingly sharp, especially given the censorship of photos on host sites. One might hypothesize that as images are less findable on a given site, the decentralized content might be more heavily searched. Rather, it appears that when established sources for the content no longer reliably provide it, those looking for it lose interest rather than dig into less recognized sources. However, it is also possible that these search queries are not for the viral content itself, but for the news media generated around it, and reflect the brief attention span of the media rather than disinterestedness among those seeking private photos.

Topics

fappening	jennifer l...	kate upt...	kim kard...	rihanna
Search term	Search term	Search term	Search term	Search term

Interest over time ⓘ

☑ News headlines   Forecast ⓘ

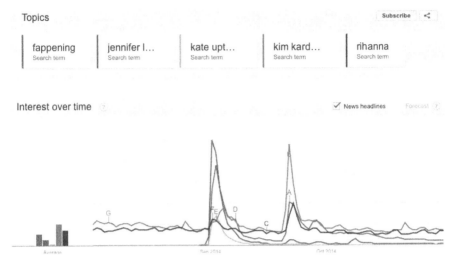

Fig. 2. Searches for "fappening", "Jennifer Lawrence photos," "Kate Upton photos," "Kim Kardashian" and "Rihanna" between August 1 and October 31, 2014.

1.2   Communicating Results

In communicating the results of this study, it would be wise to keep in mind the limitations discussed above. While Google is a powerful tool for culling information about public interest, we must be aware of what groups could be excluded. The individuals who actively avoid using the world's most popular search engine may be the very ones most interested in the illegal content released in this photo leak. Particularly given the knowledge that Google itself was removing search results (though not all search results), people looking for the content would have been even more likely to turn to other search engines or corners of the Internet. The steep drop-off in searches could indicate either a drop-off in interest or a drop-off in use of Google. While the significance of platform censorship on longevity of the viral phenomenon could be notable, using this methodology, we would only be able to speculate on a causal relationship.

In order to draw more specific conclusions about *who* was using Google to search for *what*, the study would need to employ very specific search terms. Selection of these could be based on a semantic analysis of key words in Anonymous or Reddit threads, or even a survey asking "If you were to find leaked celebrity photos online, what would you search?" If terms were not made adequately specific, the report would need to be very aware of the breadth of its terms' implications; those searching "Jennifer Lawrence" may simply be interested in her upcoming role in the *Hunger Games* trilogy. A report on broader terms could discuss "The Fappening" as a phenomenon at large more safely than it could purport to focus only on those seeking the illicit content. It would also need to be cautious about assuming that trending searches yielded the desired results, as there are at least two distinct forms of content that could be desired. In essence, a trends analysis could yield interesting visualizations of macro-level interest in terms over time, but may not target the specific population or content of greatest interest.

Communicating these results to academics would require a rigorous explanation of how search terms were selected and assurance of comprehensive comparison between them. It may also require situating the results in theoretical discourse about how our search inputs do and do not relate to our conceptualizations of the content we seek, as well as a thorough discussion of the shortcomings of our methodology. The general public may respond more strongly to visualizations of findings and a language that recognizes popular jargon for the phenomenon.

2   A Qualitative Approach

As well as generating interesting questions about the structural spread censored viral content, Celebgate ignited interesting moral issues surrounding voyeurism, responsibility, and compliance online. Is hacking a celebrity different than hacking a noncelebrity? To what extent do individuals feel accountable for viewing content that others have shared? If a hack never enters public knowledge, is it equally reprehensible? The research question that this paper will choose to pursue is: *To what extent do users of Reddit respect the morality that the site espouses as a platform, and to what extent do they disrespect it?* I have chosen to focus on Reddit as it became a predominant hosting platform for "The Fappening," and its users are fairly explicit in their discussions of the site's principles.

There are a number of effective qualitative methodologies one could employ in pursuit of this research question. This paper examines the possibility of a discourse analysis methodology, which would analyze text that users have already contributed on the site in response to policy decisions made by the site's administrators. The analysis could explore ways in which different forums on Reddit contextualize and shape debate: how is the morality of those who post on /r/beatingwomen different than those who post on /r/feminism? Which of these dialogues aligns more closely to the moral policies encouraged by the site's creators? What tone does the conversation take on in broader reaching forums, like the comments on official policy statements?

Other studies have used discourse analysis to elucidate ideological leanings of newsgroups[8] and commentary on news articles.[9] Yet in order to unearth the heart of this discourse on Reddit, one would need to examine carefully selected subreddits, rather than the random sample one might propose for a survey. It is likely that /r/freespeech would have more to say on the subject of Reddit's censorship particular subreddits than, say, /r/saxophonics. Looking at forums selected primarily for their

---

[8] Herring et al. (2004) "Women and Children Last: The Discursive Construction of Weblogs" *Into the Blogosphere*. University of Minnesota. Available Online:
http://blog.lib.umn.edu/blogosphere/women_and_children.html

[9] Diakopoulos and Naaman. (2011) "Towards quality discourse in online news comments." *CSCW*. Available Online:
http://dl.acm.org/citation.cfm?id=1958844&dl=ACM&coll=DL&CFID=648040872&CFTO KEN=10450748

ideological charge, the aim of the study would be to assess diverging or converging ideologies between platform hosts and platform users. Size of subreddit should be taken into account, as those with greater followings may have claim to broader community representation.

Specifically, this study would need to assess particular conceptions across the different forums. Firstly, we could look at the idea of responsibility: who is responsible for protecting private content from being viewed? Is it those who view it, those who share it, those who host it, those who hack it, or those who create it—the victims themselves? Secondly, we could look at the related notion of property: to whom do these photos belong and in what kind of economy should they be exchanged?

2.1   Responsibility

In the public statement that Reddit released after closing down /r/TheFappening, the site's administrators explicated their ideological stance towards hosting morally questionable materials. They wrote:

> We uphold the ideal of free speech on reddit as much as possible not because we are legally bound to, but because we believe that you—the user—has the right to choose between right and wrong, good and evil, and that it is *your responsibility* to do so. When you know something is right, you should choose to do it. But as much as possible, we will not force you to do it... Virtuous behavior is only virtuous if it is not arrived at by compulsion. This is a central idea of the community we are trying to create.[10]

This philosophy raises critical questions about the politics of voyeurism on Reddit and the location of responsibility for the diffusion of illicit content. It shoulders the position that it is not the responsibility of the platform to prevent such content from being viewed, but rather that of the site's users. However, more extreme opinions place the responsibility much farther back in the chain of dissemination. One video claiming to be an interview with the hacker responsible for the leaks states his position as such:

> I have two pieces of advice. Whenever you send a photo to someone else... Your boyfriend, friend or... Anytime, if you hit 'Send', then you should imagine that it gets sent to the whole world. Do you understand what I'm saying? Once you press 'Send' it's not yours anymore. It's not your image. It belongs to everyone.[11]

While the video is of questionable authenticity, it espouses an extreme ideology in which private communication online no longer exists and property online is collectively owned. Thus, while Reddit's official stance diffuses responsibility for

---

[10] "Every Man Is Responsible For His Own Soul" (2014, 6 September). *Blog.Reddit.* Online: http://www.redditblog.com/2014/09/every-man-is-responsible-for-his-own.html

[11] CAPSLOCK. (2014, September 3) "THE FAPPENING 2014 iCloud hacker Interview (Full Version)" *YouTube.* Online: https://www.youtube.com/watch?v=zZ_OeL2GTw8

viewing content among online voyeurs, this position places the responsibility on the victims. The discourse analysis proposed would look at conceptions of responsibility across several different forums, and draw conclusions not only about distinctions in ideology, but the relationships between those ideological discourses, the extents to which they are impacted by platform ideology, and the extents to which they actively subvert that ideology.

2.2   Property

The video mentioned above also exposes interesting arguments related to property on the Internet. Not only does it claim that the leaked photos are property of all Internet users who choose to consume them, but that they should be treated as free goods in a sharing economy. This notion dissolves typical treatment of celebrity nudes, which are traded regularly on the darknet between trusted peers in exchange for other images. This sort of direct exchange sounds like one of Lessig's hybrid economies,[12] in which nonfinancial trading supplants traditional commercial relationships. Security researcher Nik Cubrilovic states, "Most members of these rings are not financially motivated," and explains that the in order to enter a ring, you need to bring your own novel stolen material.[13] The 2014 leaks breached not only the traditional rules of how celebrity nudes are shared online, but (in allowing these images to attract media attention) the knowledge of those from whom they were stolen. Both of these breaches have notable consequences for how these images are conceptualized as property.

Numerous commenters on Reddit pointed to the forum /r/photoplunder, which self-describes as "A place to share interesting pictures of women that we find in public view."[14] The subreddits generally features non-celebrities who are almost certainly unaware that their sensitive photos are being shared at large on this forum. If victims are unaware that their images are being shared, is the leak still equally damaging? A discourse analysis might reveal differing sentiments expressed by Reddit administrators and Reddit users on the matter.

Cubrilovic finds, "It appears the intention was never to make these images public,"[15] and that some opportunist decided to take advantage of their availability. Yet to what extent are celebrities *already* viewed as public property? We share music and paparazzi photographs about celebrities openly, and claim a familiarity with them that is certainly nonmutual. As cultural icons, celebrities are arguably of enough

---

[12] Lessig, Lawrence. (2008) *Remix: Making Art and Commerce Thrive in the Hybrid Economy.* New York: Penguin. Print.

[13] Popper, Ben. (2014, September 4) "Inside the strange and seedy world where hackers trade celebrity nudes." *The Verge.* Online:
http://www.theverge.com/2014/9/4/6106363/celebgatefappening-naked-nude-celebrities-hack-hackers-trade

[14] mechesh, (2012) "Reddit: photoplunder" *Reddit.* Online:
http://www.reddit.com/r/photoplunder/

[15] Cubrilovic, Nik. (2014, September 2). "Notes on the Celebrity Data Theft" *New Web Order.* Online: https://www.nikcub.com/posts/notes-on-the-celebrity-data-theft/

public interest to consider their lives collectively owned. However, fame does not negate personhood and rights to personal property and privacy. To what extent do Reddit users and administrators treat celebrity property as collectively owned, and to what extent do they acknowledge it as private?

2.3 Possible Findings

One might expect to find varying results on all three of these questions as discourses develop in disjoint forums on the site. Cursory analysis of two different subreddits reveals highly divergent conclusions about the first question: whose responsibility is it for protecting the privacy of Fappening victim? On /r/news, one of the more mainstream forums attracting diverse respondents, discussion takes on a freedom of speech debate. In response to a post criticizing new California laws against revenge pornography, one user writes,

> You may not feel as though law enforcement has the authority to tell you what to do with nude photos of your ex-girlfriend, but surely you have a moral obligation to avoid publicly exposing and shaming another person. If that exposure makes a person feel that their right to privacy has been violated, involving law enforcement is the logical next step.[16]

This user concludes that the responsibility for preventing privacy breaches is on those who share the photo beyond their intended reach—in the case of the Celebgate leaks, that would leave the blame on the hackers who dispersed them. This view is debated heavily on /r/news, with many others arguing that the responsibility for protecting privacy should be the burden of the individual featured in the photo who shared it in the first place.

Most of the users posting on the forum /r/MensRights take a different stance on responsibility. One user argues,

> I'm not saying what should or should not be illegal. I'm just saying that to the best of my lay knowledge, if the boyfriend takes the pictures, then he owns the copyright. Period...So if an ex-boyfriend publishes nude pictures, not with the purpose of causing emotional harm, the CA law wouldn't be applicable and there would be no copyright claim.[17]

Under this argument, the responsibility for protecting privacy would fall to the victims of the leak, not those who share them. The pushback for this viewpoint is

---

[16] SlimLovin. (2014, December 3). "'Revenge' porn law: Ex-boyfriend who posted nude photos gets jail time." *Reddit: /r/news*. Comment. Online: http://www.reddit.com/r/news/comments/2o1pky/revenge_porn_law_exboyfriend_who_posted_nude/cmixy4y

[17] grocket. (2014, December 3). "Am I off base? While I agree "revenge porn" should probably be illegal, I can't help feeling like it's going to be yet another large weapon in the ammo box for false accusers." *Reddit: /r/MensRights*. Comment. Online: http://www.reddit.com/r/MensRights/comments/2o5d5j/am_i_off_base_while_i_agree_revenge_porn_should/

much less aggressive than it is on the broader site discussion on /r/news; already we can discover divergent discourses developing on different platforms. The number of subscribers to each platform should impact discussion of what can truly be considered prevailing attitudes on the site. Hopefully our results would reflect not only differences in discourse development on different forums within Reddit, but would be able to compare those fractured sentiments to the conclusions reached on more centralized platform discussions.

2.4   Communicating Results

Again, in communicating our results we must keep in mind the limitations of our methodology. A discourse analysis reveals sentiments of the more vocal members of a community, but might fail to capture a body of users who consume Reddit content without adding to it. One could perhaps ameliorate this by combining a discourse analysis with interviews, though the problems with these are, again, outlined above. It would be of critical importance to communicate these results in a way that treated a spectrum of Reddit users with dignity. Because their communications on the site are publically available and already pseudonymous, there would be little work to do in terms of obscuring their identities or acquiring consent for use—perhaps those to be included in publication could be contacted in advance via Reddit's messaging system.

Moreover, looking only at forums that tend toward ideological discussion might cause us to miss those in which users are actively not complying with the policies laid out by the site. In other words, while users on /r/MensRights might advocate against punishment of those sharing leaked photos, the real photo leaks are taking place elsewhere, on places like /r/Celebs or /r/ledzepplin. On the other hand, many subreddits featuring sensitive actively remove leaked photographs from their forums, exhibiting compliance with site policy. One could arguably expand the breadth of subreddits analyzed in order to include information on compliance or lack of compliance. However, this may require a broader conception of what constitutes a discourse and fall outside the purview of this project.

3   Conclusion

As a viral phenomenon, the celebrity photo leaks of 2014 were not the first of their kind, but they were certainly the most public. The rapidity with which they gained attention and were subsequently suppressed provokes interesting structural and ethical questions about online censorship. A variety of quantitative and qualitative methodologies are well suited to the study of different questions related to the phenomenon, and the most fruitful projects would do well to combine several.

Celebgate was a critical cultural moment, as it raised important questions about how secure we ought to feel in sharing our information, how responsible we are for its protection, and how others might go about obtaining it. Surely this heightened awareness of the risks involved in sending sensitive content could alter the extent to which that behavior manifests. For all of these reasons, these leaks are important to study; ways in which our privacy can be breached and who is responsible for protecting it are topics of the highest salience for Internet scholars and broader publics alike.

References

1. CAPSLOCK. THE FAPPENING 2014 iCloud hacker Interview (Full Version) YouTube (September 3, 2014), https://www.youtube.com/watch?v=zZ_OeL2GTw8
2. Cubrilovic, N.: Notes on the Celebrity Data Theft. New Web Order (September 2, 2014), https://www.nikcub.com/posts/notes-on-the-celebrity-data-theft/
3. D'Orazio, F.: How Stuff Spreads #1: Gangnam Style Vs. Harlem Shake—Full Study and Data Visualization. FACE blog (May 8, 2013), http://www.facegroup.com/blog/how-stuff-spreads-1-gangnam-style-vs-harlemshake.html
4. Diakopoulos, Naaman: Towards quality discourse in online news comments. CSCW (2011), http://dl.acm.org/citation.cfm?id=1958844&dl=ACM&coll=DL&CFID=648040872&CFTOKEN=10450748
5. Every Man Is Responsible For His Own Soul. Blog.Reddit (September 6, 2014), http://www.redditblog.com/2014/09/every-man-is-responsible-for-his-own.html
6. Google. Transparency Report; Request ID: 1445734. Google, (December 1, 2014), http://www.google.com/transparencyreport/removals/copyright/requests/1445734/
7. grocket. Am I off base? While I agree "revenge porn" should probably be illegal, I can't help feeling like it's going to be yet another large weapon in the ammo box for false accusers Reddit: /r/MensRights. Comment (December 3, 2014), http://www.reddit.com/r/MensRights/comments/2o5d5j/am_i_off_base_while_i_agree_revenge_porn_should/
8. Herring, et al: Women and Children Last: The Discursive Construction of Weblogs into the Blogosphere. University of Minnesota (2004), http://blog.lib.umn.edu/blogosphere/women_and_children.html
9. Lessig, L.: Remix: Making Art and Commerce Thrive in the Hybrid Economy. Penguin, New York (2008) (print)
10. mechesh, Reddit: photoplunder, Reddit (2012), http://www.reddit.com/r/photoplunder/
11. Popper, B.: Inside the strange and seedy world where hackers trade celebrity nudes. The Verge (September 4, 2014), http://www.theverge.com/2014/9/4/6106363/celebgate-fappening-naked-nudecelebrities-hack-hackers-trade
12. Shokouhi, M.: Seasonal Queries by Time-Series Analysis. In: SIGIR (2011), http://research.microsoft.com/pubs/150746/sigir0196-shokouhi.pdf
13. SlimLovin. Revenge' porn law: Ex-boyfriend who posted nude photos gets jail time. Reddit: /r/news. Comment (December 3, 2014), http://www.reddit.com/r/news/comments/2o1pky/revenge_porn_law_exboyfriend_who_posted_nude/cmixy4y
14. Stern, M.: The Fappening 4': More Celeb Nudes Leak Online Including the First Man, Seems to Be Dying Down.The Daily Beast (October 5, 2014), http://www.thedailybeast.com/articles/2014/10/05/the-fappening-4-more-celebnudes-leak-online-including-the-first-man-seem-to-be-dying-down.html
15. Vaidhyanathan, S.: The Googlization of Everything (And Why We Should Worry), pp. 2–3. University of California Press, Berkeley (2011)
16. Valdivia, Monge-Corella: Diseases Tracked by Using Google Trends, Spain. Emerg. Infect. Dis. 16(1), 168 (2010)

Internet Science and Societal Innovations

A Disciplinary Analysis of Internet Science

Clare J. Hooper[1(✉)], Bruna Neves[1], and Georgeta Bordea[2]

[1] The University of Southampton IT Innovation Centre, Southampton, UK
cjh@it-innovation.soton.ac.uk
[2] Insight, National University of Ireland, Galway, Ireland
georgeta.bordea@insight-centre.org

Abstract. Internet Science is an interdisciplinary field. Motivated by the unforeseen scale and impact of the Internet, it addresses Internet-related research questions in a holistic manner, incorporating epistemologies from a broad set of disciplines. Nonetheless, there is little empirical evidence of the levels of disciplinary representation within this field.

This paper describes an analysis of the presence of different disciplines in Internet Science based on techniques from Natural Language Processing and network analysis. Key terms from Internet Science are identified, as are nine application contexts. The results are compared with a disciplinary analysis of Web Science, showing a surprisingly low amount of overlap between these two related fields. A practical use of the results within Internet Science is described. Finally, next steps are presented that will consolidate the analysis regarding representation of less technologically-oriented disciplines within Internet Science.

Keywords: Internet Science · Disciplinary analysis · Interdisciplinarity · Bibliometrics · Natural language processing

1   Introduction

Internet Science involves interdisciplinary collaboration to deepen our understanding of the Internet as a societal and technological artefact, whose evolution is increasingly intertwined with that of human societies [1]. Like the fields of Web Science and Human-Computer Interaction [2], Internet Science's interdisciplinarity is a key strength but at times a challenge: conducting successful collaborations across disciplinary boundaries can be difficult, demonstrated by efforts to discuss [3] and address [4] [5] such issues.

This paper builds on previous work in which we established a method to empirically evaluate the presence of disciplines in an interdisciplinary field based on a Natural Language Processing (NLP) analysis of a corpus of data [6]. This method has been applied in the related field of Web Science, but here we focus on its application to Internet Science, identifying key topics and application contexts of the field.

The benefits of such an analysis are wide-ranging, relating to:

1.   **Quality**: work by Lungeanu [7] has shown that collaborator diversity is positively correlated with the quality of work.

© Springer International Publishing Switzerland 2015
T. Tiropanis et al. (Eds.): INSCI 2015, LNCS 9089, pp. 63–77, 2015.
DOI: 10.1007/978-3-319-18609-2_5

2. **Credibility**: the Internet Science community claims to be interdisciplinary [1], but at this time has little evidence to back this up.
3. **Insight**: By identifying over- and under-representation of disciplines in Internet Science, we become able to take corrective action if needed. For example, we can reach out to under-represented communities by writing targeted calls for papers and co-locating Internet Science workshops and conferences with appropriate events.
4. **Communication**: By better understanding what disciplines are active within Internet Science and what we mean by the term 'Internet Science', we can communicate better as a community. This has positive benefits for internal and external communication, as well as for curriculum planning.

The analysis in this paper is made possible by the EU Network of Excellence in Internet Science (EINS), a European-funded research instrument to coordinate and integrate European research in the area of Internet Science [1]. Among other things, EINS has fostered the formation of a group of academics actively working in this area, and this group have published widely on a variety of Internet Science topics. These publications are the key input to the analysis described in this paper.

This paper is structured as follows: We first introduce the area of bibliometrics and the use of NLP to analyse a corpus of data, and describe our previous work in this area. We then describe the disciplinary analysis method, from initial data gathering, through processing and visualising the data, to conducting a 'community analysis' of that data. We then present our results, including an analysis of application contexts within the resultant graph. After discussing these results, including a comparison of Internet Science and Web Science results as well as a discussion of how the results have been applied within current Internet Science work, we discuss avenues for future work and present our conclusions.

2   Background and Related Work

2.1   Bibliometrics, Natural Language Processing, and Disciplinary Analysis

This work uses a base assumption from bibliometric mapping [8], that a research field can be described by a list of important keywords. Previous bibliometrics work ranges from co-citation analysis [9] and examination of conferences [10] to geospatial visualisation of collaboration [11]. Previous work has typically used author assigned key phrases and pre-built domain taxonomies [12], but such resources are not readily accessible here. For this reason we apply an automatic method [13] for extraction of domain terms.

Implicit relations between the extracted topical descriptors can be discovered and described through word co-occurrence analysis, a content analysis technique that was effectively applied to analyse interactions in different scientific fields [8] [12]. This technique was applied to analyse the interconnections between a main field, i.e., fuzzy logic theory, and other computing techniques [14], a setting that is similar to our analysis of the Internet Science field. A more recent work on co-word analysis [15] outlines several limitations related to the use of keywords and proposes a method to integrate expert knowledge into the process. A main issue with this approach is that it requires a considerable amount of human intervention for the construction of domain specific thesauri. We alleviate this challenge by completely automating the process of identifying topical descriptors and by automatically constructing a domain taxonomy (topical hierarchy).

Internet Science is at the crossroad of domains as diverse as Civic Planning, Psychology and Economics. Each domain has a different level of formality, with a varying number of natural language terms and a more or less deterministic syntax. This has a direct impact on the performance of term extraction tools, with a larger number of correct terms extracted for some domains than for others. The portability of term extraction systems is rarely evaluated across domains, with most studies considering only one domain for evaluation [16] [17]. In [18], different term extraction approaches are evaluated over two domains, a biology corpus and a small general knowledge corpus of Wikipedia articles; term extraction performance is shown to vary depending on the domain. More recent work [19] studies the performance of term extraction systems over three domains (Computer Science, Biomedicine, and Food and Agriculture). That work showed that Saffron, our NLP tool, produces stable results across different domains, and for that reason we use that approach in our work studying Web Science.

This paper builds on previous work conducted by the authors in the related domain of Web Science. A WebSci'12 paper presented initial work in this area [20], and was built upon in a WebSci'13 paper [6] that analysed almost 500 articles (compared to 69 in 2012) and used a small expert survey to (1) aid our interpretations of graph structures and taxonomies and (2) dig deeper when distinguishing between disciplines. In this paper, we use those foundations as a starting point for applying the same approach in this related but distinct domain, in addition describing the practical use of the results in outputs for the Internet Science project, EINS.

2.2   Analysis of Internet Science

Since our previous publication [6], steps include an approach to visualising a given domain's literature based on co-readership [21] and, of particular note here, work to map the an EU FP7-funded Network of Excellence (NoE) [22]. This latter work examined the venues at which members of EINS published, surveying publications to include the word 'Internet' in their title, abstract or keywords; co-authorship networks were also considered. The authors found an imbalance between disciplines identified as predominant in EINS (computer science, physics) and the range of disciplines to do scholarly work about the Internet. This work differs by focusing on what disciplines are present in Internet Science publications, rather than what domains Internet Science experts previously published in.

Other work has explored Internet Science in different ways. At the first Internet Science conference, Dini and Sartori presented an excellent interdisciplinary dialogue (held between a sociologist and an engineer, both active within Internet Science) about the epistemological bases of different disciplines towards developing a method of analysis of the Internet [3]. They conclude that one unified interdisciplinary theoretical framework cannot be attained, but that it is both possible and deeply worthwhile for people with differing disciplinary perspective to collaborate towards a common goal. This is consistent with findings at a recent workshop held on overcoming barriers to interdisciplinary collaboration [5].

3   Disciplinary Analysis Method

In brief, the method involves: gathering a corpus of data; conducting Natural Language Processing to extract topics; a graph analysis and visualisation of the extracted topics; and conducting a 'community' analysis of that output.

3.1   Data Gathering Method

The corpus consisted of the set of papers downloadable from the Internet Science bibliography, online at http://www.internet-science.eu/biblio. At the time of the download, August 2014, a total of 208 papers were available. The text-processing tool we use requires that files be in PDF or TXT format. Although some of the papers at the bibliography website were not in a suitable format, most of these papers could be found using Google Scholar. In total, 207 of the 208 papers were processed.

3.2   Natural Language Processing Method

We processed the files using Saffron[1], a knowledge extraction framework for understanding research communities [23]. Saffron uses information extracted from unstructured documents using Natural Language Processing techniques [24]. An automatically constructed domain model based on the corpus of data was used, with the approach described in [19]: this domain model was used to construct several linguistic patterns that identify candidate terms based on their context. Noun phrases of maximum 5 words extracted in this way were considered as candidate terms and then ranked based on their length, frequency and embeddedness.

As described in Section 3.1, 207 files were included in the analysis. The Saffron analysis yielded 29,016 phrases that were identified as research term candidates, with an average of 140 candidates per document. The extracted terms were further used for a manual analysis of the corpus and to automatically construct a topical hierarchy. Limitations related to graph visualisation and time constraints mean our analysis considers the best ranked 1000 terms.

---

[1]   http://saffron.insight-centre.org/

The research terms are not manually curated and therefore include incorrect terms such as 'research mobility', which is not an Internet Science research term. Like any other tool, term extraction and analysis has some limitations, and the appearance of 'research mobility' as an important term exemplifies the issue of incorrectly extracted terms.

The index used in co-word analysis to measure the strength of relationships between two research terms is defined as:

$$I_{ij} = D_{ij} / (D_iD_j)$$

where D_i is number of articles that mention the term T_i in our corpus, D_j is number of articles that mention the term T_j, and D_{ij} is the number of documents in which both terms appear.

Edges are added in the research terms graph for all the pairs that appear together in at least 3 documents. Saffron uses a generality measure to direct edges from generic concepts to more specific ones. This step results in a highly dense, noisy directed graph that is further trimmed using an optimal branching algorithm. An optimal branching is a rooted tree where every node but the root has in-degree 1, and that has a maximum overall weight. This algorithm was successfully applied for the construction of domain taxonomies in [25]. This yields a tree structure where the root is the most generic term and the leaves are the most specific terms.

3.3   Graph Analysis and Visualisation Method

We used a network graph tool, Gephi, to build a graph showing links between each of the 1000 terms: nodes are extracted terms and arcs are papers that link them. This let us visually identify clusters of closely related terms. We used the Force Atlas 2 algorithm to layout the graph with the following parameters: Scaling: 2.0; Edge weight influence: 0.0. We used betweenness centrality to weight node importance. Betweenness centrality measures the fraction of shortest paths going through a node [26]: a high value indicates that nodes play an important bridging role in a network.

The lines connecting nodes in the graph represented whether two nodes appeared in the same paper together. Lines could be thicker or thinner, according to the number of times the terms appeared together (thinner lines correspond to fewer co-occurrences, while thicker lines indicate greater co-occurrence). Edges were only visualised for pairs that appeared together in at least 3 documents.

3.4   Community Analysis Method

Finally, the Louvain method [27] was applied with resolution 15 to detect 'communities', clusters of nodes that were more closely linked with one another than with the rest of the graph. We interpreted 'communities' as application contexts ranging from technologies (i.e. open hardware) to disciplines (i.e. network science) and topic areas (i.e. energy).

4   Results

In this section we first present the results of the graphing and visualisation, before moving onto the outputs of the community analysis.

4.1   Graph Analysis and Visualisation Results

Figure 1 shows a visualisation of the extracted terms, in which larger nodes and label fonts indicate terms with a higher betweenness centrality. Table 1 lists terms with a particularly high betweenness centrality, showing for comparison the top-rated terms obtained by applying the same procedure to a corpus of Web Science papers [6].

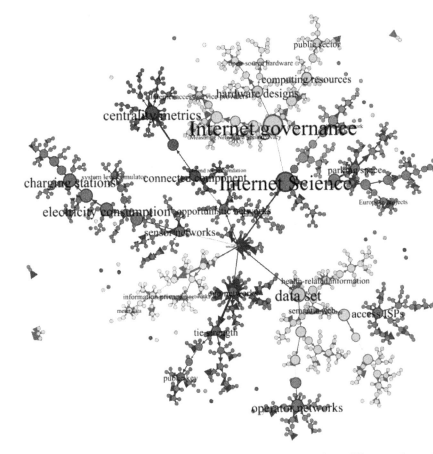

Fig. 1. Visualisation of the extracted Internet Science terms, where different colours indicate different detected 'communities' (application contexts)

4.2   Community Analysis Results

The community detection algorithm returned many 'communities' (henceforth: application contexts) with a very sparse distribution. We observed a long tail where most communities were poorly connected and constituted only a very small portion of the graph. For this reason, only application contexts representing at least 1% of the graph were considered, a total of 11/60 detected application contexts. Of the remaining 49 communities, which were very small and not linked to a substantial part of the visualised graph, many consisted of only one term.

Table 1. The 20 terms with highest betweenness centrality

Internet Science term	Internet Science Betweenness centrality	Web Science term [2013]
Internet governance	322	semantic web
Internet Science	310	social media
data set	211	information retrieval
renewable energy	192	social networking site
centrality metrics	192	social science
charging stations	185	search engine
electricity consumption	183	social networking
energy consumption	180	learning network
operator networks	165	web page
parking spots	164	personal learning environment
data collection	162	social interaction
personal data	160	mobile device
data flows	153	future research
hardware designs	147	internet user
centrality indices	147	uniform resource identifier
connected component	144	web science research
access technologies	140	user interface
access ISPs	138	web community
social interaction	138	web application
content delivery	135	linked data principle

Table 2 summarises the 11 considered application contexts, showing for each: a given name; its root node (the node with the highest betweenness centrality); the percentage of the graph that it covered; its number of 'hot terms' (terms with a betweenness centrality > 100); and its top 5 terms as ranked by betweenness centrality.

Application context names were chosen based on the topics within each application context, with a weighing towards those topics with a high betweenness centrality.

Table 2. Summary of the 11 application contexts

Name	Root node	% graph	# hot terms	Top 5 terms
Measurement technologies	data set	13.2	8	data set; operator networks; data collection; access technologies; social interaction
Energy	renewable energy	13	7	renewable energy; charging stations; electricity consumption; energy consumption; sensor networks
Social Networking	opportunistic networks	12.1	1	opportunistic networks; information privacy; social networking; online social networking; information sharing
Internet governance	internet governance	11.2	6	internet governance; personal data; data flows; content delivery; data protection
Network Science	tie strength	9.8	1	tie strength; ego networks; public key; additional information; tie strength evaluations
Parking	parking spots	8.3	5	parking spots; parking search; opportunistic networks using cognitive heuristics; parking space; cognitive heuristics
Network Science II	centrality metrics	7.3	3	centrality metrics; centrality indices; connected component; video systems; content placement
Architecture / design	internet science	6.6	4	internet science; internet architecture; virtual network mapping; network mapping; current internet
Open hardware and infrastructures	hardware designs	5	1	hardware designs; internet design; open hardware; open source hardware; software code
ISPs and Network traffic	access ISPs	4.7	1	access ISPs; transit traffic; content distribution; access ISP; traffic patterns
Research mobility	information collected	3.3	1	information collected; route selection; BGP datasets; research mobility; routing system

5   Discussion

This section is divided into four parts. Firstly, we discuss the key terms and application contexts identified in the above analysis. This is followed by a comparison of these results with the results of applying the same method to Web Science data. Thirdly, we describe the application of our results in EINS outputs, before finally evaluating our approach and discussing future work.

5.1   Internet Science Terms and Application Contexts

Our analysis yielded a ranked list of Internet Science terms and a set of application contexts. Topics reflected by the topmost terms include:

- Governance (reflected by the term 'Internet governance')
- Energy ('renewable energy', 'charging stations', 'electricity consumption', 'energy consumption')
- Network science ('centrality metrics', 'centrality indices')
- Internet Science ('Internet Science')

Terms with less clear mappings include: data set; operator networks; parking spots; data collection; personal data; data flows; hardware designs; connected component; access technologies; access ISPs; social interaction; content delivery.

As can be seen, some terms clearly map to a given topic or discipline, as in the terms mapped to network science above. Other terms, in the above paragraph, are applicable in any of a number of disciplines. 'Social interaction', for example, is relevant to disciplines including but not limited to communication, human-computer interaction, psychology, and sociology.

In addition to the ranked terms list, the 11 largest application contexts were analysed. These contexts relate to the terms: for example, the community named 'measurement technologies' contains a number of the more ambiguous terms above (data set; operator networks; data collection; access technologies; social interaction). This implies that that particular set of terms co-occurred with one another, perhaps in a particular series of papers.

Among the 11 application contexts, 'network science' appeared twice, perhaps representing two sets of network science-driven work that happened not to overlap (i.e. addressing different research questions). It is reasonable to consider 'network science' as a single application context of Internet Science, leaving 10 application contexts.

It should be noted that the 'research mobility' application context was found due to text referring to the Research Mobility aspect of EINS[2], rather than to a bona fide research topic: this term can also be discarded, leaving 9 application contexts.

The EINS introductory webpage[3] describes the focus of EINS as focused on "network engineering, computation, complexity, networking, security, mathematics, physics, sociology, game theory, economics, political sciences, humanities, and law, as

---

[2] http://www.internet-science.eu/mobility

[3] http://www.internet-science.eu/network-excellence-internet-science

well as other relevant social and life sciences." The above findings corroborate part of that statement (i.e. terms about energy and network science map clearly with areas including engineering, computation and networking), but do not provide strong evidence at this time for presence from, for example, sociology, game theory, economics, political science, humanities, or law. It is possible that NLP may be less good at showing results relating to disciplines which use more varied words (compared with arguably more repeatable terminology in technology fields), and it would be premature to conclude whether fields such as the humanities are definitely underrepresented in Internet Science or simply not highlighted by the NLP-led analysis.

5.2   Comparing Internet Science with Web Science

The precise delineation between Internet Science and Web Science is not always clear. One might argue that Web Science is a subset of Internet Science (because web technologies are a subset of Internet technologies), while at a 2013 Internet Science / Web Science workshop [28] it was argued that one could view Internet Science as a subset of Web Science. It is beyond the scope of this paper to analyse the relationship between these fields, but nonetheless of interest to provide an initial comparison of the results of applying disciplinary analysis to the two fields.

Table 1 shows the top-ranked terms from an analysis of Web Science papers alongside the top-ranked terms identified from Internet Science in this paper. What is most striking is the almost complete lack of overlap between the two lists of 20 terms: one single term, 'social interaction', is the only one to appear on both lists. Furthermore, among the two remaining sets of 19 terms, there are few thematic connections: the Internet Science terms, as discussed above, typically concern topics such as governance, energy and network science, while the Web Science terms centre on topics such as semantic web technologies, social media and information retrieval.

We may also consider the application contexts identified in Internet Science and Web Science. Although initial analysis revealed 11 Internet Science and 8 Web Science contexts, consideration of irrelevant terms such as 'research mobility' and synonymous contexts left a total of 9 Internet Science and 4 Web Science contexts, shown in Table 3.

Table 3. Internet Science and Web Science application contexts

Internet Science application contexts	Web Science application contexts [2013]
Measurement technologies	Information retrieval
Energy	Personalised learning / elearning
Social networking	Semantic web
Internet governance	Social networking
Network Science	
Parking	
Architecture / design	
Open hardware and infrastructures	
ISPs and Network traffic	

This suggests that even if Internet Science and Web Science do not have fundamental differences and overlap substantially, the two communities that were analysed here and in the 2013 Web Science work [6] do not overlap. Given that, at minimum, the two communities share similar challenges with respect to overcoming barriers of interdisciplinarity, organisations silos and the like, a greater flow of collaboration between the two is surely desirable.

We note a difference in the number of communities first detected when running the Louvain algorithm on the two datasets. In Web Science, a total of 9 contexts were initially detected, while in Internet Science, 60 were found. There is no significance to this result, which is merely due to one of the parameters of the algorithm.

We also note a difference in the number of candidate terms in each dataset. The Web Science corpus yielded 5361 candidate terms (54 per document), compared with 29,016 candidate terms (140 per document) here. This is not due to a difference between Internet Science and Web Science, but from settings in the algorithm: the first analysis used a manually defined domain model (i.e. domain-specific lexicon) for Computer Science. The analysis presented in this paper used an automatically constructed domain model derived from the corpus itself, which is more specific and adapted to the domain, hence increasing the number and coverage of terms.

5.3   Application of Results

A key EINS output within Emergence Theories and Design is an online repository of Internet Science design methods[4]. The results of the Internet Science disciplinary analysis were used to structure the repository, which is a wiki containing a list of design methods. Figure 2 illustrates the front page of the repository.

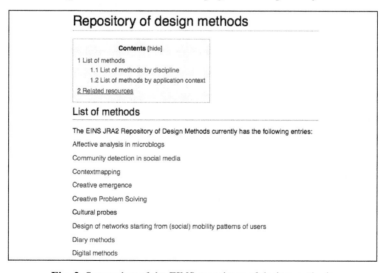

Fig. 2. Screenshot of the EINS repository of design methods

---

[4]   http://wiki.internet-science.eu/index.php/Repository_of_design_methods

At the time of writing, the repository contains 49 methods from a diversity of disciplines. A key usability goal is to make the methods accessible by classifying them in useful ways. Application contexts are a key component in that strategy, offering an ideal lens by which to structure (and help users understand) the methods.

Each repository entry consists of: a title; a free text description; a table of relevant links; a table of relevant publications; a list of applicable disciplines; a list of application contexts; any tags; a contributor. Figure 3 shows part of an entry, focusing on the later, more structured parts of the entry.

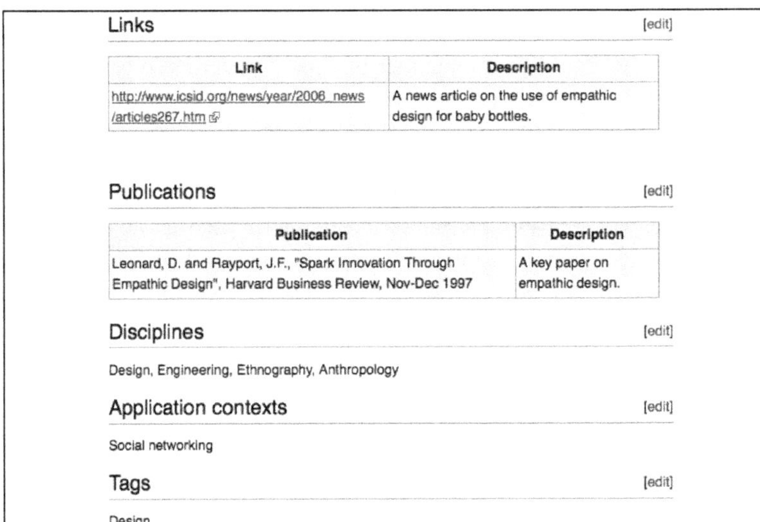

Links   [edit]

Link	Description
http://www.icsid.org/news/year/2006_news/articles267.htm ☞	A news article on the use of empathic design for baby bottles.

Publications   [edit]

Publication	Description
Leonard, D. and Rayport, J.F., "Spark Innovation Through Empathic Design", Harvard Business Review, Nov-Dec 1997	A key paper on empathic design.

Disciplines   [edit]

Design, Engineering, Ethnography, Anthropology

Application contexts   [edit]

Social networking

Tags   [edit]

Design

Fig. 3. Screenshot of part of an entry in the EINS repository of design methods

As can be seen, the method shown in Figure 3 is relevant to the application context of social networking.

Use of application methods to structure the repository offered an additional benefit: the ability to analyse the distribution of repository entries according to application context, and hence to identify under-represented areas. For more on the use of disciplinary analysis to structure and analyse the repository, see [29].

5.4   Evaluation and Future Work

As noted in Section 3.2, the terms identified by the NLP analysis are not manually curated and therefore include incorrect terms such as 'research mobility'. This exemplifies a typical limitation of approaches such as NLP, where powerful computation is used to rank terms based on factors such as frequency and co-occurrence, but there is no understanding of term meaning (and hence relevance) to the topic in question.

Section 5.1 observed that some terms clearly map to a discipline (i.e. centrality metrics with network science), but others (such as social interaction) do not. Gaining more robust insights into the link between terms and disciplines would enhance the insights gained thus far; an expert survey offers an ideal way to complement the strengths of NLP and gain such insights.

Getting input on the term-discipline link from a wide variety of experts overcomes the issues of individual assessment of such linkages, which apply particularly strongly in the context of an interdisciplinary field such as Internet Science. The issue is that two or three researchers, even ones from different fields, cannot represent sufficient disciplinary diversity to analyse and understand the implications of a set of terms sourced from highly interdisciplinary materials.

An example of applying a small survey to bolster the results of disciplinary analysis can be found in [6]: domain experts were asked for their views on what disciplines were most closely related to the top-ranked terms from the dataset. We hope to conduct a similar survey of Internet Science experts in the near future: the upcoming Internet Science conference offers an ideal opportunity to conduct this survey.

An additional benefit from the survey relates to the point in Section 5.1 that NLP may be less good at showing results from disciplines that use more varied words (such as the humanities). Internet Science survey results, combined with existing survey results from Web Science, will let us clarify whether less technologically-oriented disciplines are under-represented in Internet Science or simply less well detected by the NLP-led method used to date, providing insight not only into Internet Science but also into NLP itself.

Concerning other paths for future work, this paper has already presented a brief comparison of results from Internet Science and Web Science. Clarification of the boundaries between Web Science and Internet Science has been identified as an important area for future research [30] [31], and a full NLP-led comparison of Internet Science and Web Science materials represents a way forward in this area.

Finally, this work raises future questions regarding the use of NLP to understand differences in disciplines according to context: for example, how might sociology appear in a Internet Science corpus compared to within a pure sociology corpus? Issues include gaining datasets that are representative of a given domain: for example, the BAWE dataset at Insight has only 111 sociology documents, many of which are short student essays.

6   Conclusions

We have described the application of techniques from NLP and network analysis to gain insight into disciplinary representation within Internet Science, revealing a focus on governance, energy and network science (from terms) as well as social networking, parking, and open hardware (from application contexts). We found evidence for the presence of disciplines such as network engineering, computation and networking, and plan further analysis to investigate the presence of less technology-oriented disciplines such as sociology, humanities and law. This supplementary analysis, to be implemented via an expert survey, will also provide insight into the efficacy of NLP when working with non-technology disciplines.

In addition to providing an initial disciplinary analysis of Internet Science, we have demonstrated the applicability of our results for structuring a repository of design methods according to application contexts. We have also described a comparison of the Internet Science results with results from a disciplinary analysis of Web Science, finding a surprising lack of overlap between the two analyses, suggesting that the two communities do not strongly overlap in their current research activities.

Acknowledgements. The research leading to these results received funding in part from the EU FP7 EINS under grant agreement No 288021 and from the Science Foundation Ireland (SFI) under Grant Number SFI/12/RC/2289 (Insight). The authors gratefully thank the survey participants for their time and energy.

References

1. Network of Excellence in Internet Science, Grant agreement 288021, Description of Work (December 06, 2011)
2. Hooper, C.J., Dix, A.: Web science and human-computer interaction: forming a mutually supportive relationship. ACM Interactions 20(3), 52–57 (2013)
3. Dini, P., Sartori, L.: Science as Social Construction. In: International Conference on Internet Science, pp. 42–51 (2013)
4. Guevara, K., Blackwell, A.: A Reflective Examination of a Process for Innovation and Collaboration in Internet Science. In: International Conference on Internet Science, pp. 37–46 (2013)
5. Hooper, C.J., Millard, D.E., Azman, A.: Interdisciplinary Coups to Calamities (workshop). In: Web Science, Bloomington, Indiana, USA (2014)
6. Hooper, C.J., Bordea, G., Buitelaar, P.: Web Science and the Two (Hundred) Cultures: Representation of Disciplines Publishing in Web Science. In: ACM Web Science, Paris, France (2013)
7. Lungeanu, A.: Understanding the assembly of interdisciplinary teams. Journal of Informetrics 8(1), 59–70 (2014)
8. Callon, M., Cortial, J., Turner, W., Bauin, S.: From Translations To Problematic Networks – An Introduction To Co-Word Analysis. Social Science Information 22(2), 191–235 (1983)
9. Chen, C., Carr, L.: Trailblazing the literature of hypertext: author co-citation analysis (1989–1998). In: 10th ACM Conference on Hypertext and Hypermedia (1999)
10. Henry, N., Goodell, H., Elmqvit, N., Fekete, J.: 20 Years of 4 HCI Conferences: A Visual Exploration. International Journal of Human Computer Interaction - Reflections on Human-Computer Interaction 23(3), 239–285 (2007)
11. Nagel, T., Duval, E., Heidmann, F.: Exploring the Geospatial Network of Scientific Collaboration on a Multitouch Table. In: 22nd ACM Conference on Hypertext and Hypermedia, demo (2011)
12. Coulter, N., Monarch, I., Konda, S.: Software engineering as seen through its research literature: A study in co-word analysis. Journal of the American Society for Information Science 49(13), 1206–1223 (1998)
13. Bordea, G., Buitelaar, P.: DERIUNLP: A context based approach to automatic keyphrase extraction. In: Proceedings of the 5th International Workshop on Semantic Evaluation, Stroudsburg, PA, USA (2010)

14. Lopez-Herrera, A., Cobo, M., Herrera-Viedma, E., Herrera, F.: A bibliometric study about the reseach based on hybridating the fuzzy logic field and the other computational intelligent techniques: A visual approach. International Journal of Hybrid Intelligent Systems 7(1), 17–32 (2010)

15. Wang, Z.Y., Li, G., Li, C.Y., Li, A.: Research on the semantic-based co-word analysis. Scientometrics 90(3), 855–875 (2012)

16. Ananiadou, S.: A methodology for automatic term recognition. In: The 15th Conference on Computational Linguistics (1994)

17. Velardi, P., Sclano, F.: Termextractor: a web application to learn the common terminology of interest groups and research communities. In: 7th Conference on Terminologie et Intelligence Artificielle (2007)

18. Zhang, Z., Iria, J., Brewster, C., Ciravegna, F.: A comparative evaluation of term recognition algorithms. In: Proceedings of the Sixth International Conference on Language Resources and Evaluation, Marrakech, Morocco (2008)

19. Bordea, G., Buitelaar, P., Polajnar, T.: Domain-independent term extraction through domain modelling. In: The 10th International Conference on Terminology and Artificial Intelligence, Paris, France (2013)

20. Hooper, C., Marie, N., Kalampokis, E.: Dissecting the Butterfly: Representation of Disciplines Publishing at the Web Science Conference Series. In: Proc. ACM WebSci 2012 (2012)

21. Kraker, P., Jack, K., Schloegl, C., Trattner, C., Lindstaedt, S.: Head Start: Improving Academic Literature Search with Overview Visualizations based on Readership Statistics. In: Proc. ACM WebSci 2013 (2013)

22. Sahal, A., Wyatt, S., Passi, S., Scharnhorst, A.: Mapping EINS–An exercise in mapping the Network of Excellence in Internet Science. In: The 1st International Conference on Internet Science (2013)

23. Monaghan, F., Bordea, G., Samp, K., Buitelaar, P.: Exploring Your Research: Sprinkling some Saffron on Semantic Web Dog Food. In: Semantic Web Challenge at the International Semantic Web Conference (2010)

24. Bordea, G.: Domain adaptive extraction of topical hierarchies for Expertise Mining (Doctoral dissertation) (2013)

25. Navigli, R., Velardi, N., Faralli, S.: A graph-based algorithm for inducing lexical taxonomies from scratch. In: Proceedings of the Twenty-Second International Joint Conference on Artificial Intelligence (2011)

26. Barthélémy, M.: Betweenness centrality in large complex networks. The European Physical Journal B - Condensed Matter and Complex Systems 38(4), 163–168 (2004)

27. Blondel, V.D., Guillaume, J.-L., Lambiotte, R., Lefebvre, E.: Fast unfolding of communities in large networks. Journal of Statistical Mechanics: Theory and Experiment 2008(10), P10008 (2008)

28. Fdidia, S., Tiropanis, T., Brown, I., Marsden, C., Salamatian, K.: 1st International Workshop on Internet Science and Web Science Synergies. In: ACM Web Science 2013, Paris, France (2013)

29. Hooper, C.J. (ed.): EINS Deliverable 2.1.2: Repository of methodologies, design tools and use cases. Network of Excellence in Internet Science FP7-288021 (2014)

30. Hooper, C.J., Hedge, N., Hutchison, D., Papadimitrious, D., Passarella, A., Sourlas, V., Wuchner, P.: EINS Deliverable 2.3: Whitepaper on recommendations for funding agencies. Network of Excellence in Internet Science FP7-288021 (2014)

31. Stavrakakis, I., Hutchison, D.: EINS Deliverable 13.2.1: Internet Science - Going Forward: Internet Science Roadmap (Preliminary Version). Network of Excellence in Internet Science FP7-288021 (2014)

Analysing an Academic Field through the Lenses of Internet Science: Digital Humanities as a Virtual Community

Alkim Almila Akdag Salah$^{(\boxtimes)}$, Andrea Scharnhorst, and Sally Wyatt

eHumanities Group, Royal Netherlands Academy of Arts and Sciences,
Amsterdam, The Netherlands
almila.akdagsalah@ehumanities.knaw.nl
http://www.ehumanities.nl

Abstract. Digital Humanities (DH) has been depicted as an innovative engine for humanities, as a challenge for Data Science, and as an area where libraries, archives and providers of e-research infrastructures join forces with research pioneers. However DH is defined, one thing is certain: DH is a new community which manifests and identifies itself via the Internet and social media. In this paper we propose to describe DH as a virtual community (VC), and discuss the implications of such an epistemic approach. We start with a (re)inspection of the scholarly discourse about VCs, and the analytic frameworks which have been applied to study them. We discuss the aspects that are highlighted by taking such a stance, and use the guidelines proposed by the FP7 European Network of Excellence in Internet Science (EINS) in our investigation.

Keywords: Virtual communities · Emergence of new scientific fields · Network analysis · Digital humanities · Bibliometrics

1 Introduction

The concept of a *Virtual Community* (VC), just like the concept of *community*, defies a generic definition. On the Internet, social links are formed through capabilities created in the online medium (e.g. blogs, shared online platforms that allow comments and messages, hyperlink references, etc.), thus bringing together people and organizations with shared interests. In this paper, we use approaches developed to investigate virtual communities to analyse Digital Humanities, treating it as a virtual -scientific- community.

There are different links between the study of the emergence and evolution of VC and the study of scientific communities. Both rely on work about the formation of social communities, the role of communication (networks) in this process and the institutional and economic base of human activity leading to the formation of communities.

© Springer International Publishing Switzerland 2015
T. Tiropanis et al. (Eds.): INSCI 2015, LNCS 9089, pp. 78–89, 2015.
DOI: 10.1007/978-3-319-18609-2_6

Scientific communities, from the *Republic of Letters*[1] to the modern *invisible colleges* [1], can be seen as the archetype of virtual communities. They existed even before the emergence of the Internet, and they are well studied. The *Republic of Letters* is itself an object of study in the Digital Humanities, which sometimes also appears to model itself after this early global network of knowledge exchange.

In this paper, we ask to what extent current research on Virtual Communities can help us to better conceptualize and analyze Digital Humanities, which from the very beginning has embraced digital environments, and with them an active online life. Which aspects of the emergence and continuous functioning of virtual communities then are relevant when reflecting about the current state and future course of DH?

In the next two sections we briefly describe both objects (VC and DH, respectively) using techniques of information retrieval, bibliometrics and visualization. In the case of virtual communities, we list important topics and analytic dimensions in the form of a baseline matrix. As we will argue, Digital Humanities scholars show similar behavioral patterns to the members of a generic virtual community. With this argument in mind, the question we highlight throughout this paper is, how can research on virtual communities be beneficial in analyzing Digital Humanities as an academic discipline?

2   Mapping Virtual Communities

One of the oldest and best known definition of virtual communities comes from Rheingold in 1993: "[all] social aggregations that emerge from the Net when enough people carry on those public discussions long enough, with sufficient human feelings, to form webs of personal relationships in cyberspace" [2]. At that time, the Internet was not part of everyday life and did not offer more than a fleeting platform to build a community (amongst the academics and computer professionals who had access to it). Some scholars questioned whether the internet would destroy community building altogether, rather than generating spaces for new communities to emerge, or if the new communities offer the same intimacy of offline communities [3].

A first exploration of the term *Virtual Communities* with a new interactive visual interface *Ariadne*[2] reveals many journals whose textual space resonates with the search term [4] (see Fig. 1). Many of these are situated at the interface between computer sciences, information sciences, sociology and economics.

---

[1] The 'Republic of Letters refers to the community of scholars that emerged in the late 17th century, who developed and maintained long-distance correspondence and exchange of ideas. To see different projects that use computational techniques to map the relationships between people and ideas check e.g. 'Mapping the Republic of Letters. See: http://www.culturesofknowledge.org/ and http://ckcc.huygens.knaw.nl/epistolarium/

[2] Ariadne (thoth.pic.nl/relate?) allows to surf through context of search terms in the *ArticleFirst* database of OCLC.

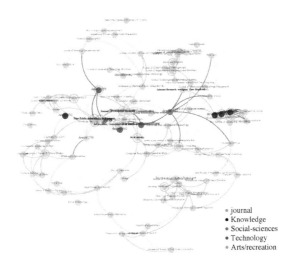

Fig. 1. Contextual exploration of *Virtual Communities*

A search in the bibliographic database *Web of Knowledge*[3] retrieves 1418 documents. In a global map of science [5] those documents are situated mostly in various branches of the Computer Sciences, Information and Library Science, Management and Business, Education and Communication. The papers citing this core set follow the same distribution, except here, we see also publications from Psychology and Cognitive Sciences (see Fig.2 (b)).

If we take a closer look at the scholarly publications around the term, we see that in the early years of the Internet, the definitions mostly were grounded in the usage of certain tools: Email, chat rooms such as Internet Relay Chat (IRC), discussion forums, etc. Access to such tools of course required a certain socio-economic background (to own a computer, to have access to the Internet, to be technologically literate enough to use these tools, etc.) which could be excused for at least a general combining factor for these loosely defined communities. However, probably a closer look at their shared interests, goals, and needs would inevitably crumble these communities into subgroups. These first attempts can be called the first wave in Virtual Communities research. The second wave, seeing the shortcomings of basing community formation on technological advancement only, extended the definition to focus on the process not the technological infrastructure. Thus, virtual communities started to be explored around the ideas of shared benefits and goals [6]. The most recent wave brings the economic aspects of virtual communities to the fore [7].

A manual inspection of documents from this information retrieval exercise led to a number of prevailing topics such as PROTOCOLS (these constitute

---

[3] TOPIC: ("virtual communities") Timespan: All years, Web of Science Core collection, DOCUMENT TYPES: (ARTICLE OR REVIEW OR BOOK OR PROCEEDINGS PAPER OR EDITORIAL MATERIAL OR BOOK CHAPTER), Feb 25, 2015.

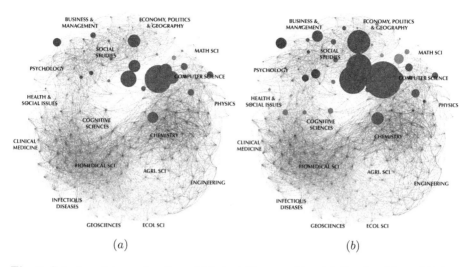

Fig. 2. Scientometric exploration of *Virtual Communities*. (a) Overlay Map of VC as a topic. (b) Overlay Map of all citing papers VC as a topic.

the background of virtual communities such as infrastructures, platforms, tools, software as well as norms and standards of the web), MEANS (this topic focuses on the perspective of the user: needs, goals, interests, i.e. human emotions that holds the community together), and CURRENCIES (these are more the economical factors: investment in social capital, time, and the expected benefits from participating in a virtual community). In Fig. 3 these topics are arranged on the left side. JRA6 developed another categorization of topics relevant to the study of virtual communities, shown at the right side of the same figure. These are socio-technical requirements (community members' social and technological needs, and requirements that arise from the interaction of social and technical domains), Virtuality (the connection and interactions between online and offline world(s)), and topologies (e.g. features of interfaces and type of activities, such as gaming or writing). The two sides of the table, i.e. Protocols, Means and Currencies and Socio-technical requirements, Virtuality and Topology of VC interact on several levels with each other. Through these interactions the shape of the virtual communities remains in a constant transformation.

3   Digital Humanities

Digital Humanities still struggles in its self-perception and definition [8]. Similar to other emerging or (relatively) newly emerged fields in humanities (i.e. Visual Cultural Studies), the community is still very much engaged with questions [9] such as: What is Digital Humanities? What are its main methods? and more importantly, What is its object of analysis?

Even the origins of Digital Humanities is open to debate: some proponents suggest the pioneering work of priest Roberto Busa, who lemmatized the works

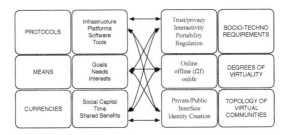

Fig. 3. A baseline matrix for *Virtual Communities*

of Saint Thomas Aquinas even before computers were invented, as the starting point of DH [10]. Others have argued that a connection between humaniora and other sciences actually was the norm in times of natural philosophy, and that Digital Humanities is now the bridge between the 'two cultures' that grew apart with the emergence of disciplines in the 19th and 20th centuries [11].

Digital Humanities (DH) are often defined as research and teaching activities which combine computing and information technologies with scholarly practices in humanities fields. But, current aspirations of DH go far beyond methodological innovation. A White Paper written to support the foundation of a new center lists as challenges: "1) transformational progress in humanities research and understanding to address societal challenges, 2) significant improvements in algorithms and computational instruments that deal with heterogeneous, complex, and social data, and 3) societal benefits through novel understandings of language, culture, and history." (p.8) [12]. With such aspirations the group of possible members of the community is large: scholars in humanities fields and in computer sciences, those involved in education, science and education policy makers, and basically every citizen.

A first exploration again with *Ariadne* reveals another active community: libraries and archives. The map in Fig. 4 shows a group of library journals at 12 o'clock, and then clockwise: computer & information science journals (e.g. about ontologies); education and various specific humanities areas (e.g., architecture, visual arts, literary history and archaeology). The heterogeneity of off-line communities concerned with Digital Humanities is also visible in a bibliometric analysis of its formal scholarly communication.

The query "Digital Humanities" in the Web of Science[4] retrieves a set of 390 papers. All the articles citing this set is another set of 281 articles[5]. First of all,

---

[4] TOPIC: ("digital humanities" or "humanities computing" or "e-humanities" or "computational humanities") Timespan: All years, Web of Science Core collection, DOCUMENT TYPES: (ARTICLE OR REVIEW OR BOOK OR PROCEEDINGS PAPER OR EDITORIAL MATERIAL OR BOOK CHAPTER), Feb 25, 2015.

[5] Total Citing Articles (without self-citations) of the TOPIC: ("digital humanities" or "humanities computing") Timespan: All years, Web of Science Core collection, DOCUMENT TYPES: (ARTICLE OR REVIEW OR BOOK OR PROCEEDINGS PAPER OR EDITORIAL MATERIAL OR BOOK CHAPTER), Feb 25, 2015.

this is a relatively small number, even one recognizes that Arts & Humanities are not well-represented in this database and that the citation culture of the humanities prefers books to articles as the primary publication venue. The field is extremely small, and so a traditional bibliometric analysis produces rather a restricted map.

Fig. 4. Contextual exploration of *Digital Humanities*

Earlier bibliometric studies of Digital Humanities have already shown that citation analysis reveals a limited perspective of its activities [13,14,15]. For example, even though one of the main funding (and hence research) for DH is in digitization projects, and software, platform and tool building, these activities do not have a visible publication venue in the community [13]. Moreover, a deeper analysis of DH by extracting data not only from ISI, but also from Google Scholar with its broader coverage, shows that the community is focused on "on collaboration, institutional forms, library & information science, and a more limited visibility for substantive fields, including education, history, linguistics" [15]. Fig. 5 shows the overlay maps for Digital Humanities. They support these general findings: the community seems to be publishing in fields such as Computer Science and Information Studies mainly. The humanities disciplines which surface in the overlay maps are Literature, Linguistics, Educational Research and History. The overwhelming presence of Computer Science and related fields also suggests that when humanities scholars involved with DH research publish in their home disciplines they do not use DH as a keyword in their publications. To see how widely DH has infiltrated humanities citation literature, it may be useful to use a combination of webometrics and scientometrics (where the names of every humanities scholar who claims to do research on DH is collected, and used as the starting point for data collection from ISI).

In the data collected for overlay map analysis, a handful of publications are listed as "Humanities, multidisciplinary", and this categorization veils the fine-

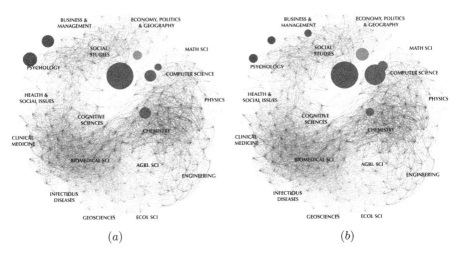

(a) $\qquad\qquad\qquad\qquad\qquad\qquad\quad$ (b)

Fig. 5. Scientometric exploration of *Digital Humanities*. (a) Overlay Map of DH as a topic. (b) Overlay Map of all citing papers DH as a topic.

grained details of which humanities disciplines actually collaborate for Digital Humanities research. To see the international and the institutional collaboration of Digital Humanities, the same data sets are combined and an analysis is run with the tools developed by Loet Leydesdorff [16]. The resulting networks clearly captures the main players in the field: UK and USA as the main collaborators, followed by the usual suspects in Europe: Holland, Finland, Sweden, France, Germany. Fig. 6 (a) shows unexpected names to pop-up as well, for example nations geographically not connected to Europe are already connected to the main players: Kuwait, Tanzania, South Korea. Fig. 6 (b) renders the interdisciplinary nature of DH with interconnected but small groups, hence a main component and many smaller cliques dominate the visualization.

It is not by chance that one of the best-known mapping exercise about DH - the infographic *Quantifying Digital Humanities*[6] contains a lot of altmetrics, including:

– subscriptions to the Humanist discussion list
– number of people on Twitter identified as Digital Humanities scholars
– registered users of thatcamp.org - a platform to organize conferences in the field.

This infographic was published in 2012. Since then, DH has grown, and now exhibits many of the characteristics of an emergent field which has left the stage of prematurity and lonely pioneers and moved even beyond the stage of early adopters. The field has its own professional organization, local and international

---

[6] http://melissaterras.blogspot.nl/2012/01/infographic-quanitifying-digital.html

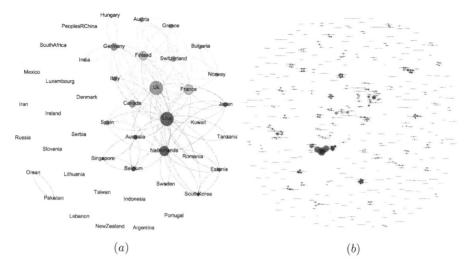

Fig. 6. Collaboration analysis of *Digital Humanities*. (a) International collaborations. (b) Institutional collaborations.

conference series, dedicated journals, specific funding programmes, and an increase in DH minors, master and PhD programmes.

DH has fully embraced all forms of digital scholarship [17]. However, this is not the only feature of a virtual community. What about the connection and tension between off-line and on-line identity; the relevance of the physical; the importance of access to sustainable and adaptive infrastructures; and most importantly, what about the goals, needs, interests of its members?

4   Digital Humanities as Virtual Community

There are many online outlets that are regularly used by academicians: email, library/publication databases and their information retrieval systems, personal websites, course websites, teaching platforms such as Moodle, Blackboard, and massive online courses (MOOCs). The academy generally provides the infrastructure and necessary platforms for these online outlets, covering certain issues that the JR6 pointed out in its reports: trust/privacy, user interfaces, operability, durability, public/private, etc. On the other hand, there are many tools/platforms that are not generated for academic use, but used by some scholars: blogging, podcasting, Twitter, Facebook, Flickr (as a visual data repository and sharing), Youtube, Dropbox, etc. Then, there are platforms specifically designed for (collaborative) research: Mendeley, Academia.edu, Researchgate, etc. Mundane and specific digital activities are continuously intertwined.

Digital Humanities, among this plethora of possibilities, not only make the most from infrastructure, platforms, tools and software, but DH scholars are also actively involved in building all these for their special needs and purposes[7]. They also keenly advocate for e-publication and open data policies. A possible change in the current standards of evaluation in academia to include diverse online activity is also an important point discussed in DH. This discussion takes place under the heading of how to reward for other, often invisible labor [18], such as professionally kept blogging sites, or developing and sharing new curricula. DH tries to expand the boundaries of academical infrastructure, poses questions, experiments with different practices, and helps evolving new services, platforms and tools. Hence, to a large extent DH operates as a VC, and/or is a part of a non-academic VC. The Network of Excellence EINS (JRA6 in particular) has emphasized several dimensions which are important in a user-centric analysis of virtual communities, including socio-technical requirements; virtuality and topology [19]. In the following, we use those dimensions to discuss current affairs in Digital Humanities.

The Role of Socio-technical Requirements and User Needs. One could possibly write the history of DH as a history of tools and platforms that were developed and used. Some were incorporated into daily practices and emergent infrastructures; others were project-specific and faded away. It is quite evident that with the on-going technological evolution tools used in DH will continue to change. One continuing problem in DH has been and still is the need to build protocols and norms, and to stop generating similar tools and platforms. A proper information flow between funding agencies, and already finished projects is a much needed, but still missed resource for the community. Another need is in the dissemination and teaching of the developed tools and platforms. Workshops on how to apply tools will remain important, but what is also needed is to reach a basic level of computer literacy across all humanities disciplines. Course designs which equip users with skills that enables them to adapt to new developments will become more important than learning a specific tool which might be outdated within a year. Standardization, i.e. the identification and consolidation of those techniques generic to many humanities fields, is another need counterbalancing the ongoing innovation of methods, techniques and tools. Next to machine-supported text analysis, such standardization also concerns other type of material: audio, visual (static, i.e. images or dynamic, i.e. videos), numeric information and information-ordering techniques as harmonization and mapping between ontologies. All these activities (sometimes doubling an already finished project) fall under the umbrella of PROTOCOLS, as much as the requirements that arise from socio-technical needs.

Degrees of Virtuality. One important aspect of social communities in the Internet age is that they are never only on-line or off-line, there will be always hybrid

---

[7] A generous amount of funding for Digital Humanities projects are reserved for digitization of Humanities archives, as well as to build platforms so that such digitization projects will be durable, and available for the wider scholarly community and public at large

forms. Scientific communities are no exception to this. In an envisioned continuum between virtual and real, different activities of members of the DH community can be placed at different locations. From the point of view of MEANS, i.e. user needs, goals and interest, scientific communities such as DH has already a binding focus offline, which is nurtured via standard academic outlets such as common publication venues, conferences, workshops etc. DH designs many of these outlets in forms of online communications as well, i.e. attending a conference via a podcast, preferring e-publications and blogging for scholarly communication, or using online data-repositories and platforms for collaboration are normal occurrences within the community. But continuing from the perspective of future user needs, there clearly is a need for not only Virtual Research Environments, but physical centers. In times of project funding and competition, temporary centers, groups, network organizations are rather the norm than the exception. But a sustainable development of a new emerging field needs some stable institutional form. The establishment of chairs at universities might be one way to add more stability. There is also another lesson learned from VC, which is a plea for off-line counter parts. In contrast to off-line communities, on-line communities require more often a renewal of membership by activities. This is a burden on limited time resources, and one of the reasons that VC cease to function. Without an off-line counterpart, a physicality, one cannot expect persistence and continuation [20].

Topology of Virtual Communities. Under topology we understand the infrastructures and the content which is created using them. Also in this case, project-based funding might counter consolidation. Excellent for exploration, short- and mid-term funding instruments are counterproductive when it comes to the maintenance of infrastructures, and to the consolidation of knowledge. The sheer size of current information spaces in academia entails the danger of a *balkanization* of knowledge, and the wasting of resources in imitation and redundancy, instead of using them for knowledge discovery and innovation. For DH there are many information sources on-line and off-line, but there is - as for almost all fields - also a great need for registries, catalogues, interactive maps to resources, in short intelligent information management. Next to this stands the question of access to those topologies. Access to literature but also access to servers.

5   Implications and Future Work

One way to build continuity in VC is the generation of norms and rules. For scientific communities this was already articulated in the 1960s by Robert Merton and others. However, it is equally important to note that while we can righteously talk about norms, values and an academic culture, we should not forget that this culture relies on infrastructure, not only technical, but also centers for education and research, and all funding mechanisms which help these centers to thrive. Here, the struggle of DH for further growth and maturity is part of the overall battle of the humanities against marginalization[8].

So far we have reflected about the current stage of DH using categories relevant for VC. But what the above considerations also reveal is that we have little empirical evidence for good arguments towards funding bodies to further support and establish DH - as virtual and as real-world community. Mapping of newly emerging fields happens only occasionally, and we are far monitoring their birth more regularly, and check if they are part of sciences, economy or society. We have seen in the case of DH that the well-established monitoring tools, based on bibliometrics of formal scholarly communication, fail to capture essential features. Having said this, VC, because of their on-line habitat, reveal *in principle* possibilities of monitoring. Altmetrics has already paved this path. But, one can also see that altmetrics starts from what is measurable, and these might not always be the features one really would like to monitor in the emergence of a new field. What Digital Humanities need most is not altmetrics, but a good management of information flow. DH should function as they would be a VC, and not as an enterprise distributed across numerous academic fields at once.

References

1. de Solla Price, D.J., Beaver, D.: Collaboration in an invisible college. American Psychologist 21(11), 1011 (1966)
2. Rheingold, H.: The virtual community: Homesteading on the electronic frontier. Perseus Books (1993)
3. Wellman, B., Gulia, M.: Net surfers don't ride alone: Virtual communities as communities. Networks in the Global Village, 331–366 (1999)
4. Koopman, R., Wang, S., Scharnhorst, A., Englebienne, G.: Ariadne's thread: Interactive navigation in a world of networked information. In: CHI 2015 Extended Abstracts (2015)
5. Rafols, I., Porter, A.L., Leydesdorff, L.: Science overlay maps: A new tool for research policy and library management. Journal of the American Society for information Science and Technology 61(9), 1871–1887 (2010)
6. Chiu, C.M., Hsu, M.H., Wang, E.T.: Understanding knowledge sharing in virtual communities: An integration of social capital and social cognitive theories. Decision Support Systems 42(3), 1872–1888 (2006)
7. Stokburger-Sauer, N.E., Wiertz, C.: Online consumption communities: An introduction. Psychology & Marketing 32(3), 235–239 (2015)
8. Svensson, P.: The landscape of digital humanities. Digital Humanities Quarterly 4 (2010)
9. Terras, M., Nyhan, J., Vanhoutte, E.: Defining digital humanities: A reader. Ashgate Publishing, Ltd. (2013)
10. Busa, R.A.: Foreword: Perspectives on the digital humanities. In: Schreibman, S., Siemens, R., Unsworth, J. (eds.) A Companion to Digital Humanities, pp. 187–188. Blackwell Publishing Inc. (2004)
11. Bod, R.: A new history of the humanities: The search for principles and patterns from antiquity to the present. Oxford University Press (2014)

---

[8] http://4humanities.org/

12. Wyatt, S., Millen, D. (eds.): Meaning and Perspectives in the Digital Humanities. Royal Netherlands Academy of Arts and Sciences (2014)
13. Akdag Salah, A.A., Scharnhorst, A., Leydesdorff, L.: Mapping the flow of digital humanities. In: Digital Humanities Conference (DH 2010). Kings College, London (2010)
14. Leydesdorff, L., Salah, A.A.A.: Maps on the basis of the arts & humanities citation index: The journals leonardo and art journal versus digital humanities as a topic. Journal of the American Society for information Science and Technology 61(4), 787–801 (2010)
15. Wyatt, S., Leydesdorff, L.: e-humanities or digital humanities: Is that the question? In: Digital Humanities Workshop (2013)
16. Leydesdorff, L., Wagner, C.S.: International collaboration in science and the formation of a core group. Journal of Informetrics 2(4), 317–325 (2008)
17. Borgman, C.L.: Scholarship in the digital age. MIT Press (2007)
18. Wouters, P., Beaulieu, A., Scharnhorst, A., Wyatt, S.: Virtual knowledge: experimenting in the humanities and the social sciences. MIT Press (2013)
19. Marsden, C., David-Barrett, T., Pavan, E., Ciurcine, M., Arata, G., Mantelero, A., Brown, B., McMillan, D., Cave, J., Trossen, D., Pierson, J., Talboom, S., Passarella, A., Antoniadis, P., Rouncefield, M., Karaliopoulos, M.: Deliverable 6.1: Overview of user needs analysis, plus draft catalogue of design responses to needs analysis (2013)
20. McCarty, W.: Humanities computing: essential problems, experimental practice. Literary and Linguistic Computing 17(1), 103–125 (2002)

EINS Evidence Base: A Semantic Catalogue for Internet Experimentation and Measurement

Xin Wang[1(✉)], Thanasis G. Papaioannou[2], Thanassis Tiropanis[1], and Federico Morando[3]

[1] Web and Internet Science Group
Electronics and Computer Science
University of Southampton, Southampton, UK
`xwang@soton.ac.uk, t.tiropanis@southampton.ac.uk`
`http://www.ecs.soton.ac.uk/`
[2] Information Technologies Institute
Center for Research and Technology Hellas (CERTH), Thessaloniki, Greece
`thanasis.papaioannou@iti.gr`
`http://www.iti.gr/`
[3] Nexa Center for Internet & Society
Politecnico di Torino, Turin, Italy
`federico.morando@polito.it`
`http://nexa.polito.it/`

Abstract. To explore the socio-technical aspects of the Internet requires infrastructures to properly foster interdisciplinary work and the development of appropriate research methods. To this end we present a platform called EINS Evidence Base (EINS-EB) which is developed as part of the EINS project. The EINS-EB also aims to empower researchers, academics, organisations and society to engage with Internet Science research independent of background. Currently, it provides for the collection and discovery of data resources, of analytic and simulation tools, and, in the future, of the methodologies behind those tools an of relevant scholarly activity. We explore issues of data representation, dataset description, dataset catalogues and method catalogues for Internet Science. The evidence base adopts semantic technologies to provide an interoperable catalogue of online resources related to Internet science. We also present activities on making the evidence base interoperable with related e-Science activities by communities engaging in relevant interdisciplinary collaboration.

Keywords: Evidence base · Schema.org · Web observatory · Semantic catalogue

1 Introduction

The Internet Science community has been actively engaged in intensive research with respect to networking parameters issues to performance, QoS, security and availability guarantees. As this discipline evolves, Internet Science has become an

© Springer International Publishing Switzerland 2015
T. Tiropanis et al. (Eds.): INSCI 2015, LNCS 9089, pp. 90–99, 2015.
DOI: 10.1007/978-3-319-18609-2_7

"inter-discipline" that draws on disciplines such as computer science, economics, sociology and law, and requires to go beyond the networking techniques to the exploration of correlations between the evolution of the Internet infrastructure and of the societal and business sectors of activity that increasingly rely on it. To this end there is demand to set up e-Science infrastructures that are appropriate to foster research collaboration among disciplines, empowering the communities involved with access to methods, data resources, data collection tools, analytic tools, and simulation tools. Many of those tools are available but they scattered in different repositories maintained by different communities making it hard to discover and use them especially by members of different disciplines.

An appropriate representation of resources in the repository is essential in order to transform mere data into information about the structure, function and performance of the system. In addition, the ability to capture all (potentially) relevant aspects of the broader context, repeat an experiment, reuse the same data and combine various types of information for analytic purposes is vital to establishing and communicating the robustness, generalisability and implications of particular findings in line with accepted scientific methods. On the other hand, an expressive representation usually comes at the cost of complexity. A main challenge is to adopt and extend a representation that is expressive enough to capture relevant characteristics of resources without introducing too much complexity. We decided to use the Schema.org vocabulary which is a lightweight, flexible and extensible vocabulary that is supported by main search engines. We adopt and extend the Schema.org vocabulary and build a Internet Science resource repository called the EINS Evidence Base (EINS-EB). Furthermore we deploy semantic enrichment technologies to assist users to record rich metadata of published resources with minimum efforts.

The remain of this paper is organised as following: we provide an overview of the EINS-EB in Section 2; we discuss the design concerns and describe the techniques of the EINS-EB in Section 3; details of data hosting and the infrastructure of EINS-EB are given in Section 4; we discuss the relationships and differences between the evidence base and several related platforms in Section 5, and the conclusion and future plan in Section 6.

2 Overview of the EINS-EB

The aim of the evidence base is to foster interdisciplinary work by creating an online catalogue that will record and expose detailed metadata of existing selected datasets, methodologies and tools. Interoperability of the representation of catalogued resources is essential for the evidence base to be used by researchers form different disciplines, and a well designed community engagement mechanism is crucial for gathering as many as possible resources that are related to Internet science. With those requirements in mind, the EINS-EB is built to enable users to conveniently publish and share Internet science resources with rich semantics.

In the EINS-EB, there is emphasis on open datasets (as shown in Figure 1) to ensure engagement with the wider community around Internet science. Apart from datasets, the online resource will catalogue related tools that are required to collect, analyse, visualise data, and e-infrastructures which are necessary to carry out Internet science experiments in controlled environments.

For each resource catalogued in the EINS-EB, general information such as title, description, keywords etc. are recorded to enable prompt searching. Based on the textual description of each resource, a DBpedia[1] classification is automatically generated by the system (and can be further specified or corrected by the publisher). The evidence base does not facilitate direct access to listed resources, but it requires publishers to provide URLs from where resources are available, along with licensing information. All the above mentioned information is embedded as Microdata in the web pages using Schema.org vocabularies. Notice that pages are marked up using the simple Microdata syntax and the Schema.org vocabulary, as explained below; nevertheless, a mapping from Schema.org to RDF (expressed in RDF Schema) is available and applications can use services to obtain a Linked Data representation of the Microdata. This makes our approach simple and pragmatic, yet essentially compatible with the more ambitious efforts to build the Web of Data.

Name	Licensing	Keywords	DBpedia	Description
Stanford Large Network Dataset Collection	Open	social networks, ground-truth community networks, communication networks	Social Networks	This is a collection of datasets (collected by the...
Eurosys '09	Proprietary	Communications, Facebook, Social graph	Social_networks	Datasets containing Facebook social graph (friends...
WOSN '09	Proprietary	Social networking service, Facebook, Social network	Social_networks, Social_graph	Dataset containing Facebook social graph (friendsh...
MPI-SWS Datasets	Proprietary	Social network graph, Social interactions	Social_networks, Social_graph	Social network datasets (Flickr, LiveJournal, Orku...
M-Lab data	CC0	Net neutrality, traffic discrimination , network performance	Network performance	Measurement Lab (M-Lab) is an open, distributed se...

Fig. 1. The dataset view of the EINS evidence base

It is possible for anyone to create an account and share resources on the evidence base. Listing a resource can be easily done by filling of form containing required information as stated above. To reduce user efforts, the evidence base provides a list of common licenses from which users can choose the appropriate ones. The evidence base also automatically analyses the descriptions and if possible the "About" page of published resources to provide suggestions of

---

[1] www.dbpedia.org/

DBpedia classification. More of this automatically classification is given later in Section 3.3. User account information is used to fill in publisher information of resources.

3   Cataloguing Resources with Rich Semantics

To provide a semantic-rich catalogue for Internet science related resources a key factor is to have the appropriate vocabulary to express the metadata of resources. Datasets take a significant proportion of these resources and therefore the choice of vocabulary is biased to capture as many as possible characteristics of datasets. At the same time, the chosen vocabulary also needs the ability to be extended to cover other resources such as tools and e-infrastructures.

The expressivity of a vocabulary usually comes at the cost of complexity. To reduce user efforts of providing accurate metadata, we employ a service called TellMeFirst which assists users with candidate classifications of resources by analysing their descriptions.

3.1   Microdata

An emerging approach supported by the dominant search providers is to use Microdata (http://schema.org) markup and vocabularies to describe Internet Science datasets available online. Many sites are generated from structured data, which is often stored in databases. When this data is formatted into HTML, it becomes very difficult to recover the original structured data. Schema.org is a collection of schemas that webmasters can use to markup HTML pages to describe the data structure in ways recognized by major search providers, and that can also be used for structured data interoperability (e.g. in JSON). Search engines that support Microdata including Bing, Google, Yahoo! and Yandex. On-page markup enables search engines to understand the semantics of the information on web pages and provide richer search results in order to make it easier for users to find relevant information on the web. Markup can also enable new tools and applications that make use of the structure.

Microdata is a simple semantic markup scheme that is an alternative to RDFa and it has been developed by WHATWG.The Microdata effort has two parts: markup and a set of vocabularies. The vocabularies are controlled and hosted at Schema.org. The markup is similar to RDFa in that it provides a way to identify subjects, types, properties and objects. The sanctioned vocabularies are found at Schema.org and include a small number of very useful ones: people, movies, etc. When a taxonomy for the description of the various properties of online resources is missing, then one can use DBpedia (http://en.wikipedia.org/wiki/Wikipedia:Quick_cat_index) and other widely adopted taxonomies.

The Microdata markup consists of three basic tags: itemscope, itemtype, itemprop. An itemscope attribute identifies a content subtree that is the subject about which we want to say something. The itemtype attribute specifies the subjects type. An itemprop attribute gives a property of that type. As an example, observe the embedded tags in the HTML markup of Figure 2 describing a dataset collection.

```
<div itemscope itemtype="http://schema.org/Dataset>
  <a href="http://snap.stanford.edu/data/" itemprop="name">
    Stanford Large Network Dataset Collection</a>
  <meta itemprop="http://schema.org/url"
    content="http://snap.stanford.edu/data/">
</div>
```

Fig. 2. An example of Microdata using a Schema.org vocabulary. It describes a dataset with the name *Stanford Large Network Dataset Collection* and the URL *http://snap.stanford.edu/data/*.

3.2   The Schema of the Evidence Base

Regarding online resource description, we employed Schema.org, as it is a solution capable of offering high searchability and simplicity. We chose to use at least those properties and vocabularies from Schema.org corresponding to the ones specified in the standard Dublin Core (http://dublincore.org/documents/dces/). We briefly describe the main choices regarding our schemas for the online datasets, online tools and e-infrastructures below.

Dataset We partially adopted the type `Thing::CreativeWork::Dataset` from the Schema.org vocabularies for describing the various datasets available online. This type inherits some interesting attributes (named "Properties") from the `Thing` type that can be utilized for describing the various Internet Science datasets found, depicted in Table 1.

Tool and e-Infrastructure Schema. We employed a subset of the type `Thing::CreativeWork::SoftwareApplication` from the Schema.org vocabularies for describing the various Internet tools and eInfrastructures available online.

We again employ here the same attributes inherited from the types `Thing` and `CreativeWork` that were employed for the description of an online dataset, described in Table 1. In Table 2, we briefly describe additional attributes from types `CreativeWork` and `SoftwareApplication` that are employed in the schema of the online tools and e-Infrastructures.

3.3   Description-Based Automated Classification

TellMeFirst[2] [2] is a tool for classifying and enriching textual documents via Linked Open Data. TellMeFirst leverages natural language processing and

---
[2] http://tellmefirst.polito.it/

Table 1. The Dataset schema

Attribute	Description
(inherited from type **Thing***)*	
description	Text description of the dataset at the original site.
sameAs	URL link to the original site of the dataset or the wikipedia entry describing the datasets nature.
url	The URL of the dataset.
additionalType	Categorizes the type of the dataset using alternative vocabularies or taxonomies than Schema.org ones, e.g. typeof Dbpedia categories, etc.
(inherited from type **CreativeWork***)*	
author	It may coincide with the "creator" attribute below, and refers to the creator of the dataset.
copyrightHolder	It refers to the license of the dataset.
copyrightYear	The year of the license.
datePublished	The date that the dataset became available online.
keywords	keywords describing the dataset. These are many times given in the web page of the dataset, but more "standards" classification keywords, e.g., from ACM taxonomy http://www.acm.org/about/class/2012, can be used.
audience	The scientific community of the dataset.
creator	It may coincide with the "author" attribute above.
dateCreated	The date of the dataset creation.
dateModified	The date of the dataset update.
version	This attribute can be used in case that there are multiple versions of the dataset available.
(inherited from type **Dataset***)*	
catalog	A data catalog which contains a dataset.
distribution	A downloadable form of this dataset, at a specific location, in a specific format.
spatial	The range of spatial applicability of a dataset, e.g., for a dataset on EU demographics, EU.
temporal	The range of temporal applicability of a dataset.

Semantic Web technologies to extract main topics from texts in the form of DBpedia resources. Input texts may then be enhanced with new information and contents retrieved from the Web (images, videos, maps, news) concerning those topics.

3.4   Linking to Other Semantic Catalogues

Using TellMeFirst, the resources described in the evidence base may be tagged using a very broad vocabulary (consisting of the more than 4.6 million of entries of the English Wikipedia), which is also structured by the Wikipedia community, so that entities (i.e., pages) are nested within categories, which are in turn sitting within a three of higher level categories.

Table 2. The schema for online tools and eInfrastructures

Attribute	Description
(attributes inherited from types Thing *and* CreativeWork *listed in Table 1)*	
...	
(additional attributes inherited from type CreativeWork*)*	
audience	The intended audience of the item, i.e. the group for whom the item was created.
citation	A citation or reference to another creative work, such as another publication, web page, scholarly article, etc.
contributor	A secondary contributor to the CreativeWork.
provider	The organization or agency that is providing the service.
sourceOrganization	The Organization on whose behalf the creator was working.
version	The version of the CreativeWork embodied by a specified resource.
(inherited from type SoftwareApplication*)*	
applicationCategory	Type of software application, e.g. "Traffic Generator, Network Simulator".
downloadUrl	If the file can be downloaded, URL to download the binary.
featureList	Features or modules provided by this application (and possibly required by other applications).
fileFormat	MIME format of the binary (e.g. application/zip).
fileSize	Size of the application / package (e.g. 18MB). In the absence of a unit (MB, KB etc.), KB will be assumed.
installUrl	URL at which the app may be installed, if different from the URL of the item.
memoryRequirements	Minimum memory requirements.
operatingSystem	Operating systems supported (Windows 7, OSX 10.6, Android 1.6).
processorRequirements	Processor architecture required to run the application (e.g. IA64).
releaseNotes	Description of what changed in this version.
requirements	Component dependency requirements for the tool.
softwareVersion	Version of the software instance.

The link with Wikipedia/DBpedia is also a gateway to the Web of Data: in fact, DBpedia is at the centre of the Linked Open Data Cloud[3] and linking to this core resource indirectly generates semantic relations with many other resources.

Furthermore, the Schema.org vocabulary used by the EINS-EB is compatible with many other vocabularies, and it is straightforward to import metadata from other semantic catalogues that use one of the compatible vocabularies. For example, the Southampton University Web Observatory (SUWO) [5,4] also adopts Schema.org. EINS-EB and SUWO can crawl each other's pages and list each other's resources. CKAN[4] provides DCAT documents about datasets which

---

[3] http://lod-cloud.net/
[4] http://ckan.org/

are also compatible with the Schema.org vocabulary used in the EINS-EB, and also can be imported to the EINS-EB. This interoperability virtually leads to a global network of semantic catalogues and enables users to search with rich semantics resources from any catalogue in the network.

4 Data Hosting and Infrastructure

A complementary service provided by the evidence base is data hosting. The hosting service (as shown in Figure 3) is supported by various infrastructures such as Hadoop Distributed File System (HDFS), MongoDB, SQL database etc. Users can upload data into any of supported databases and provide a link on the EINS-EB. The hosting service provides extra flexibility for sharing and reusing data. For example, the Neubot[5] data consist of a large number of files which can only be analysed after downloading them. By utilising the hosting service the Neubot data are also served by MongoDB at the University of Southampton from where users can query the data online.

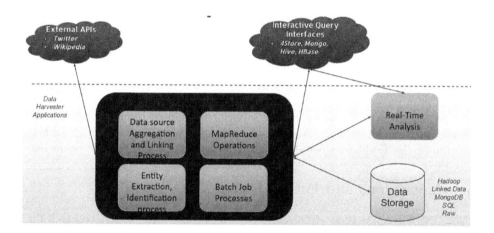

Fig. 3. Data hosting infrastructure of the EINS evidence base

5 Related Work

The EINS-EB is aiming at harmonisation with a number of similar infrastructures and tools in the areas of measurement collection, analytics and data repositories. This section outlines those related efforts and how the EINS-EB aims to interoperate with them.

Measurement Lab (M-Lab)[6] is the largest repository of open Internet performance data and has been backed many Internet measurement publications (e.g.

---

[5] http://neubot.org/
[6] http://www.measurementlab.net/about

[3,1]). It hosts a large collection of open source Internet measurement tools, data collected by those tools and visualisations based on the data. Comparing to M-Lab, the EISN evidence base focuses on providing rich metadata (via Schema.org vocabularies and TellMeFirst service) for registered resources and improving interoperability to other disciplines. M-Lab resources can be listed on the evidence base to gain better discoverability.

Southampton University Web Observatory (SUWO) [5,4] is a portal for datasets and analytics. It provides metadata of registered resources as well as facilitates access to those resources in a secure way. Resources listed on SUWO can be private and permission has to be required before accessing them. SUWO also gives Schema.org Microdata, and thus the EINS-EB can import resources from SUWO, and add extra classification information using TellMeFirst.

DatCat[7] is an online catalogue of datasets regarding Internet measurements. They employ a proprietary list of object types to describe the data collections, having each object comprising multiple fields, and maintain indices that allow advanced data querying through their web site. However, their datasets are not searchable through search engines, as opposed to our approach due to Schema.org Microdata descriptions. Moreover, their vocabulary does not adhere or link to any standard for describing data semantics.

CKAN[8] is an open data catalogue platform widely used by several data hubs. CKAN focuses on publishing metadata of datasets to increase discoverability, while the EINS-EB fosters not only datasets but also other resources that are related to Internet science, such as tools and e-infrastructures. CKAN provides a Linked Data presentation of the metadata it catalogued while the evidence base utilises Schema.org Microdata, which is better recognised by major web search engines such as Google, Bing, Yahoo!, and Yandex. Furthermore, the EINS-EB is assisted by the TellMeFirst service to automatically classify registered resources.

6 Conclusions and Future Plan

The EINS Evidence Base (EINS-EB) is fostering interdisciplinary research for the Internet Science community by providing, at first instance, a way to catalogue and discoverer related data resources and analysis or simulation tools available online. Requirements for harmonisation with existing repositories have led us to technological choices for resource description (Schema.org) and a number of utilities based on linked data and content analysis.

Leveraging the fact that Schema.org Microdata can be easily represented as Linked Data expressed with the RDF formalism (as mentioned above), we could couple the existing Microdata syntax with a formal and explicit RDF representation. This can easily be done within the HTML of the evidence base pages, using the RDFa serialisation, but with some additional effort dereferenceable IRIs can also be provided, as well as a SPARQL end-point. In this way, our initial pragmatic (and general purpose search-engine oriented) approach can be maintained,

---

[7] http://imdc.datcat.org
[8] http://ckan.org/

while the interoperability with Linked Open Data catalogues, including CKAN, may be increased. Note that TellMeFirst related data are already expressed as DBpedia resources, so connecting these to the Linked Open Data cloud would be trivial.

References

1. Basso, S., Meo, M., Martin, J.C.D.: Strengthening Measurements from the Edges: Application-level Packet Loss Rate Estimation. In: ACM SIGCOMM Computer Communication Review 2013 (2013)
2. Futia, G., Cairo, F., Morando, F., Leschiutta, L.: Exploiting Linked Data and Natural Language Processing for the Classification of Political Speech. In: International Conference for E-Democracy and Open Government 2014 (2014)
3. Masala, E., Servetti, A., Basso, S., De Martin, J.C.: Challenges and issues on collecting and analyzing large volumes of network data measurements. In: Catania, B., Cerquitelli, T., Chiusano, S., Guerrini, G., Kämpf, M., Kemper, A., Novikov, B., Palpanas, T., Pokorny, J., Vakali, A. (eds.) New Trends in Databases and Information Systems. AISC, vol. 241, pp. 203–212. Springer, Heidelberg (2014)
4. Tiropanis, T., Hall, W., Hendler, J., de Larrinaga, C.: The Web Observatory: A Middle Layer for Broad Data. In: Big Data, pp. 129–133 (2014)
5. Tiropanis, T., Hall, W., Shadbolt, N., De Roure, D., Contractor, N., Hendler, J.: The Web Science Observatory. IEEE Intelligent Systems 28(2), 100–104 (2013), http://eprints.soton.ac.uk/354604/1/TheWebScienceObservatory-postprint.pdf, http://ieeexplore.ieee.org/lpdocs/epic03/wrapper.htm?arnumber=6547975

Living with Listening Services: Privacy and Control in IoT

Donald McMillan[1(✉)] and Antoine Loriette[2]

[1]Mobile Life Centre, Stockholm University, Stockholm, Sweden
don@mobilelifecentre.org
[2]School of Computing Science, University of Glasgow, Glasgow, UK
a.loriette.1@research.gla.ac.uk

Abstract. In this paper we discuss the impact to home, work, and civil life from the deployment of continuous listening services. An example service we call the Continuous Speech Stream (CSS), would provide a real time list of keywords generated from the user's spoken interactions with others. Based on a user-study that engaged 10 users to record a full day of audio for processing into a sample stream, we report the concerns expressed by our participants on being misrepresented by their speech, unintentionally sharing sensitive data, and being unable to curate their presentation of self. We offer an initial set of recommendations for the design, testing, and deployment of IoT based services built on such rich, personal data.

1 Introduction

The success of commercial speech interfaces – such as Apple's Siri, Google's Now, and Microsoft's Cortana – has highlighted the increasing appetite for speech as in input to systems and services. Increasingly such systems work by leveraging the ubiquitous network connections of such devices to reduce power consumption and the need for processing power on the device itself by sending the audio over the internet to be analysed. This has lead to some concern. LG's latest smart TVs [1] included a feature which allowed the user to command the television at any time, without needing to indicate that the system should start to listen. This means that in order for this feature to work the TV would have to listen continuously. The TV did not process the audio itself, it was then sent over the Internet to a third party that analysed the audio and returned the results for the OS to act upon. The audio clips themselves are not only destined for algorithmic analysis, which adds another level of concern. Recent articles in the news [14] have focused on the perceived breach of privacy in having people employed to listen to the clips sent via Apple and Microsoft to evaluate and improve the performance of these algorithms.

The number of services that are based upon listening to their surroundings continuously in order to recognise speech is increasing. Amazon's Echo and The Ubi (theubi.com) not only signpost the acceptability of such listening services to consumers, but highlight the need for research in understanding the implications of their deployment. Most of these current commercial speech systems focus on a

T. Tiropanis et al. (Eds.): INSCI 2015, LNCS 9089, pp. 100–109, 2015.
DOI: 10.1007/978-3-319-18609-2_8

dialogic model of interaction, where users enter a dialogue with a system, with speech results recognised for immediate system activity, often with spoken responses, and the possibility for further dialogue. Yet recent advances in battery life, bandwidth, and processing power of mobile devices as well as the move towards multiple situated, connected sensing appliances offers the possibility of services that continually listen to natural conversation. With the opportunities of technologies such as smart watches or other wearable devices (such as the Motorola Hint earpiece), clear audio recording of everyday conversation is also easier to foresee. Technically, this offers opportunities for systems that could modify and change their behaviour based on a users' implicit spoken interaction with others. For example, phones could automatically carry out actions based on spoken audio – searches could be opportunistically run and displayed on lock screens, adverts could be better based on user activities, and models of user behaviour could make use of richer descriptions of users' lives. Socially, professionally, and civically however the impact of having all of our speech recorded and available to third parties needs to be examined.

2   Exploring Listening Services

Speech has been used as an input in HCI in a number of ways. The most common are dialogic systems where the user interacts with the system as part of a conversation, such as Siri. These are now commonly available on consumer devices. There has been research on improving the interface for speech dictation, by making it more immediate and interactive [8] and controlling a traditional touch and click based UI using speech input has also been tried in systems such as SpeechAct [24].

Using speech as a secondary input for information retrieval, for example automatic transcription and indexing of voice mails [22] and meetings [21] has received considerable research. Beyond transcription, problems such as topic extraction, detection of conversational 'hot spots', and recording speaker decisions have been addressed [20]. Systems where the speech is used in the background are rare, however one such example is the Ambient Spotlight [6] where supporting documents were retrieved to match the current topic in an ongoing meeting. It is also possible to perform tasks such as 'mining' sociometric data [23] to understand the social connections of subjects. This has been done with continual listening on mobile devices to extrapolate the audio levels and sound characteristics of social activity and to detect patterns of social interactions between users [3].

Building on work that process the audio into text, and annotating it with other features, it is also possible to employ natural language processing to analyse the text extracted. There are many approaches that can be used, in particular text summarisation approaches, covered in detail by Lloret & Palomar [11], which support extraction of key phrases and words from text. One simple approach is to extract the *subject* term from sentences. This can be done with a fairly high accuracy – over 90%.

The security implications of the systems outlined in this space come from two directions. The first being that the processing capacity required for accurate and quick transformation of audio to text or sociometric data combined with the desired mobility of the service means that there is a strong requirement to use a cloud based solution. The difficulties of securing cloud-based and IoT sensing services have been discussed by Kozlov et al [7]. That the pre-processed data will be transferred over the Internet for processing and stored by the services provider, with the results returned by the same channel opens up a number of security problems related to third parties, but there are other security concerns relating to the relationship between the user and the service provider and the split in hardware, infrastructure, and service control.

The other driver for the security problems is the use of audio itself. A microphone is not a discriminating sensor, meaning that those in the surroundings will be recorded as well as the users. It is also an incredibly rich and detailed source of data. While we focus on the spoken words that users say in range of the microphone the data collected can be used for a number of activity recognition purposes [10], and a persons voice itself is a biometrically significant indicator in identifying an individual [5].

In exploring the acceptability, impact, and opportunities surrounding always listening services we recruited 10 participants to record a day of their lives and be interviewed with regards to the Continuous Speech Stream (CSS) we manually produced. Participants wore a lapel microphone mounted onto their clothing, with small digital audio recorders worn on the belt or in a pocket. For privacy reasons we allowed participants to stop and start the recorder at will, and also offered to delete any sections from the audio that the participants did not want to share with us. This provided us with a small but rich dataset of naturalistically recorded speech. Participants ranged from an architect, an elementary school teacher, a brand consultant, a special needs carer and a mystery shopper. All participants were requested to inform those around them that they were recording and to wear the microphone openly during the day. Four participants recorded a day during which they were not working, six participants recorded a weekday comprising of work and after work.

The service we envisioned would combine automatic speech recognition and natural language processing. The first stage would be to use ASR to provide a transcript of the spoken words and phrases. The quality of the audio, volume of the speech and training on a particular person's voice has a great impact on the accuracy of the automatic speech recognition. Taking this into account we decided that only the owner of the device would have their speech transcribed, background conversations and conversation partners would be ignored in their stream. This restrictive setup is arguably the most likely to be rolled out in a commercial product, using for example the affordance of ear bone microphone, by circumventing the problem of noisy input.

We determined that natural language processing would be used to extract *subjects* and *objects* of speech, filtering out personal pronouns, for inclusion in the stream. We also decided to include *time phrases* and *numbers* wherever they appeared grammatically given their utility. Such processing of an ASR transcript we deemed to be computationally feasible. We note that the attributes of the ASR and the NLP stages could be tuned to include or exclude a number of attributes – grammatical, aural or statistical – however this simple definition of the CSS was compatible with both our goals for design and for privacy. Not providing a full transcription, not promising accuracy of recognition, not providing timestamps – all of these were to introduce ambiguity, uncertainty and therefor deniability into the system. This produced a word list from the day for each participant, covering all the topics spoken about over this period. Overall, the corpus of word transcribed formed a dictionary of 2500 distinctive words, with a flat distribution and top 10 words being uttered 5 to 15 times a day. Participants completed a semi-structured, transcribed interview in which their word list was used as a prompt.

3   Impact of the Continuous Speech Stream as a Service

In our initial design exploration of the possibilities afforded by such a service available on mobile devices [12] we determined possibilities at a personal level (such as reflecting on ones stream), on a shared level (such as subscribing to another's stream for a period of time), and on an aggregate level (such as providing the sentiment of an area by counting the occurrences of a particular topic of conversation). Here we look at the different aspects of daily life that would be affected by the widespread adoption of continuous listening services.

3.1   Home Life

One of the most straightforward uses of this information would be as a memory aids. The amount of conversation taking place, the topics (work related or personal), and the counts of certain words can be used for a number of self-reflective purposes. However the knowledge that such word counts and topics are being recorded, and that they would be available to share if one allows adds another dynamic to the conversational interactions taking place within the home. As with social media, the accountability of sharing and hiding can cause friction, and will continue to do so until the social conventions and expectations around its use settle. With more complex analysis of the streams relationships could be examined for their quantitative and qualitative attributes, such as the common topics of conversation, positivity of language used, and their similarities to other relationships, as well as the duration and frequency of interactions. This seemingly objective data on the on-going state of relationships within the home could cause problems – indeed, paraphrasing Schoeman [19], one view of privacy has at its core the right to your unadulterated self-image. Your personal perception of your closest relationships would necessarily be affected by such as system, and even without sharing the information it could be seen in this sense as a severe privacy violation.

3.2   Work Life

Another of the service designs put forward was that of providing people awareness of your current situation through sharing the CSS. Actively sharing of a length of time in the style of the Glympse application for location sharing (glympse.com), a user could share a meeting with colleagues who are not present. Active reading, where the participants of a meeting could allow their streams to be accessible during that time, would provide the same service but with a different social dynamic. This would allow those arriving later to avoid topics that have been discussed, or prepare for what is currently under discussion in the room as they are arriving.

Friends and family, or famous people, could in this way be subscribed to in order to increase awareness and connectedness, with simple filters set up to alert when two people are talking about the same topic or say each other's name, for example. But in a work setting management could, as has been the case with Facebook and other social networking sites in the past, demand access to the streams of their employees. It would also be technically fairly simple to install listening software on corporately controlled machines, providing good coverage of the working environment. While this could be metricised to provide feedback on the mood in the office it could also be used to enforce productivity – talking about anything but work related topics could be quantified per employee, as could talking negatively about the company or product.

In the same way as we outlined above in the home life scenario, sharing a stream that is by design an inaccurate snapshot of oneself could be even more problematic in a less casual environment. Being held accountable for a distorted transcript of one's conduct in the workplace could have serious consequences, especially if these transcripts are stored in perpetuity by an employer used as part of a formal evaluation process or as a basis for something that could be seen as a 'ground truth' in letters of recommendation to future employers.

3.3   Civic Life

One of the most interesting possibilities for even a small percentage deployment of such a system would be the ability to extract information on the topics of conversation in an area to form a new type of demographic. For advertisers this would lead to a new metric, the "talkthrough". It would allow those serving adverts to be alerted to the number of situations where keywords from their advertisement, be they the brand name or a other salient feature of the product, have entered the conversation of the viewer. In [12] we discussed the pattern of conversation around pre-purchase routines that would allow advertisements to be specifically targeted to those who are in the market for a competing product and to track if that consumer considered, talked about, or even bought the alternative.

But while this could be seen as a natural extension of Google's advertiser metric services this could also be employed by political parties. The action of polling an area or demographic for a sentiment could be augmented by allowing polling agencies to purchase the number of occurrences of certain words or phrases in a specific area or within a specific demographic group over a period of time. This could be used to measure statistics such as the dissemination of marketing campaigns or the political engagement of voters on a particular topic.

Of course these are, in some sense, personal surveillance systems. The use of the distributed system not to help pollsters and advertisers, but to alert government authorities to the people and places involved when 'subversive' discourse is taking place is easily envisaged. It is even possible that such technologies are already used by security services. For example, the electronic privacy information centre used a freedom of information request to obtain a list of 'trigger words' that the US department of homeland security uses as signs of terrorism. There are also organisations, such as Stratign (statigen.com), that sell GSM intercept systems that both break GSM encryption on calls and watch for particular 'trigger phrases' in speech or text. The importance of these devices to law enforcement has been highlighted by reports that the authorities would rather drop prosecution on a range of offences rather than expose how they work and what information is being collected about suspects and bystanders [18].

Even without these sinister uses there is cause for concern on the use of these services by polling agencies and political parties. As covered in the review of research on the quantification of society by Espeland and Stevens [4], the use of metrics to drive policy has changed the cause and effect of societal change. As correlations are found with data science and acted upon with greater conviction the rate of this change is increasing. In light of this, providing a shortcut to seemingly objective truth of political sentiment could greatly empower the demographics feeding data into the system at the expense of the technologically or economically excluded.

4   Participants' Concerns

Our self-selected trial participants voiced some discomfort with being recorded at all times, and that in itself suggests that such continuous listening services may be slow to reach wide acceptance. Confidential information being inadvertently disclosed during the use of this system was one such concern, with one of our participants modifying his speech to not include the name of a particular client.

"I didn't say the name, because it is, the client is a secret client, but [co-worker] said it all the time."

There was also some concern as to the persona that would be projected by a transcript of everything they say in a day.

"I was a little apprehensive, because I thought, Oh God, everyone's going to know how stupid I am all the time."

But most of our participants reported some degree of normalisation to the process; with half of them reporting forgetting the recording at least for part of the time.

There is considerable ambiguity in the keyword stream we produced, so the potential for a user to be held accountable for a particular topic of conversation was lower than it may first appear. This, of course, is taking the continuous speech stream as an isolated data point. When combined with other sources of data it provides yet another point of triangulation and in combination with location data and time could add enough context to a keyword to strongly suggest an action. The other side of the ambiguity of the stream is the potential for topics, or even errors in recognition, to be taken out of context causing concern and embarrassment:

"Porn [...] I think we were perhaps joking about something, but I can't remember exactly [...] I really wouldn't want to share that word with family."

The image presented by participants through their CSS was also explored during the interviews, showing that the without the context surrounding the words spoken the readers preconceptions are strongly projected onto the producer of the stream.

"This must be a girl that right here.x It's the first person. It looked like it's a person that is a bit stressed about the age because talking about Botox all the time and her legs."

This participant was in fact reading about a long-running joke between one the stream producer and a co-worker about being the youngest workers in the office – yet without the context, tone of voice, facial expression, and body language the conclusion that the producer is vain and older than she is was drawn.

Some of these concerns could be to do with the novelty of such close and continuous surveillance, however it has been shown that these concerns can drop over time – even among those opposed to such surveillance – as familiarity increases [16].

5   Recommendations

While the continuous logging of sensor and usage data, such as location, browser history, or app usage is becoming commonplace a users' speech and contextual audio is more personal and rich resource. Not only must the lessons learned in giving users graduated control of there exposure to logged data be applied in this area, the underlying implications that the device, and through it the service provider, is continuously listening must be dealt with honestly and openly if there is to be any chance of these technologies being accepted. That the audio clips Siri and Cortana's speech recognition systems found hard to understand are listened to by office workers to produce an improved training set is not unexpected [14] – yet the implication that your audio is more likely to be listened to by a human being in situations where you are not interacting deliberately with the system, or when distracted from the interaction by a third party, is one that should be made clear to users. The idea of control and empowerment of users through ensuring their understanding of logged location data and use data has been positioned [15], and expanding this work to provide a better understand both users' current understanding of their aural privacy, and how we can design to empower them when it is threatened by emerging technology is an active area of research.

We recommend that the user should be given mechanisms to understand and intervene in the output of their processed audio data. In helping mitigate the risk of 'porn' being shared as a topic of conversation as described above a system could offer users control through word, concept, or context filtering, or by obscuring new or unexpected content in some way until it is approved. This should include giving users the option to edit to the data they have produced in the system, with the consequences (such as reduced accuracy in the future, or inability to contribute to certain aggregate measurements) clearly explained. In many ways this control of the self projected through the system mirrors the controls on social media – yet there is an important difference between having the ability to selectively share and edit carefully content carefully curated by the user and content that is automatically generated about the user.

Control of the input to such a system is more complicated. Google gives access to all the audio clips you have sent to their cloud service for transcription; yet how useful is a list of time-stamped audio files for the task of managing your privacy? Indicators of how this data has been processed (such as what features are extracted, stored, and transmitted), how it will be processed (if it is flagged for human verification, or to be used in another system), and control over the future storage and processing should be commonplace for such personal data.

RRI (Responsible Research and Innovation) has become an increasingly important field of research in Europe [17] may be affected by potential outcomes from prototypes and technology demonstrators. By moving away from the bio-medical based ethical framework taken as a starting point in many human centred technology fields (HCI [13], for example) RRI seeks to provide a more accessible, flexible, and responsive way to ensure ethical conduct while fostering innovation that has societal impact and has at its heart the goal to change the basis of ethics in innovation to include the members of society who stand to be affected by the future implications of research prototypes and technology demonstrators [2]. By applying such a method to the design and development of continuous listening services the concerns and goals of users, those incidentally captured, representatives of law enforcement, and representatives of those with business interests in such a system would be taken into account at the research design phase.

The acceptability of such services to the general public will not be a uniform. As we can see from Westin's privacy indexes [9] the concern people feel for certain types of privacy sensitive sharing of data changes by personal outlook and is changed by events. In applying any form of RRI to the design and development of systems and services based on continuous listening devices stakeholders from across the range of these privacy indexes – High (Fundamentalist), Medium (Pragmatist), and Low (Unconcerned) – should be involved.

These recommendations should be seen as a starting point for the investigation into how users' autonomy can be ensured when using such services, and research into securing transfer, storage, and processing of sensor based IoT services should be taken into account. Involving representatives from society in the design and development of systems that have the potential to touch every aspect of our lives is one way to not only demonstrate that ethical and responsible research is being undertaken, but to stay ahead of at least some of the potential problems of deploying such pervasive systems.

6   Conclusion

We can see that the addition of continuous listening services could have a profound impact on all aspects of a users life, and impact those around them even if they are not active users of such a service. As what it means to be recorded, and the meaning of the metrics extracted from such recordings, are slowly understood by individuals and society the impact of such services will change and settle – as with all socially disruptive technologies.

Coming back to one of Schoeman's definitions of privacy including autonomy of self image, while objective information on the state of your exercising and dieting provided by health applications may intrude on this self-image it can be argued that they do so objectively, for your own good. The same is hard to argue for a system that counts the number of affectionate terms said to a particular person per week or compares the number of topics discussed by one couple against a demographically calculated average. The autonomy of the personal reflection on the state of a relationship is interfered with, but as yet there is no objective improvement or goal that can be offered. This is a generalizable problem with systems and services that move from exposing the objectively quantifiable towards the more qualitative aspects of life. Presenting information to users without understanding it fully, or without providing the tools for the users to understand and contextualize the data and the limits of its relevance could be harmful to the users in question, their friends and colleagues, and the IoT sector as a whole.

References

1. Not in front of the telly: Warning over 'listening' TV, BBC News, http://www.bbc.co.uk/news/technology-31296188 (accessed on February 25, 2015)
2. Bernd, C.S., Eden, G., et al.: Responsible research and innovation in infor-mation and communication technology: Identifying and engaging with the ethical implications of ICTs. Responsible Innovation 199 (1988)
3. Chon, Y., Lane, N.D., et al.: Understanding the coverage and scalability of place-centric crowdsensing. In: Proceedings of the 2013 ACM International Joint Conference on Pervasive and Ubiquitous Computing, pp. 3–12. ACM (2013)
4. Espeland, W.N., Stevens, M.L.: A Sociology of Quantification. European Journal of Sociology / Archives Européennes de Sociologie 49(03), 401–436 (2008)
5. Fazel, A., Chakrabartty, S.: An overview of statistical pattern recognition techniques for speaker verification. IEEE Circuits and Systems Magazine 11(2), 62–81 (2011)
6. Kilgour, J., Carletta, J., et al.: The Ambient Spotlight: Queryless desktop search from meeting speech. In: Proceedings of the 2010 International Workshop on Searching Spontaneous Conversational Speech, pp. 49–52. ACM (2010)
7. Kozlov, D., Veijalainen, J., et al.: Security and privacy threats in IoT architec-tures. In: Proceedings of the 7th International Conference on Body Area Networks, pp. 256–262. ICST (Institute for Computer Sciences, Social-Informatics and Tele-communications Engineering) (2012)
8. Kumar, A., Paek, T., et al.: Voice typing: a new speech interaction model for dictation on touchscreen devices. In: Proceedings of the SIGCHI Conference on Human Factors in Computing Systems, pp. 2277–2286. ACM (2012)
9. Kumaraguru, P., Cranor, L.F.: Privacy indexes: a survey of Westin's studies (2005)
10. Lane, N.D., Miluzzo, E., et al.: A survey of mobile phone sensing. IEEE Communications Magazine 48(9), 140–150 (2010)
11. Lloret, E., Palomar, M.: Text summarisation in progress: a literature re-view. Artificial Intelligence Review 37(1), 1–41 (2012)

12. McMillan, D., Loriette, A., et al.: Repurposing Conversation: Experiments with the Continuous Speech Stream. In: Proceedings of the 33rd Annual ACM Conference on Human Factors in Computing Systems. ACM (2015)

13. McMillan, D., Morrison, A., et al.: Categorised ethical guidelines for large scale mobile HCI. In: Proceedings of the SIGCHI Conference on Human Factors in Computing Systems, pp. 1853–1862. ACM (2013)

14. Morris, I.: Apple's Siri And Microsoft's Cortana Record Your Voice, And Someone Is Listening, Forbes.com, http://forbes.com/2015/02/24/apples-siri-and-microsofts-cortana-record-your-voice-and-someone-is-listening/ (accessed on February 25, 2015)

15. Morrison, A., McMillan, D., et al.: Improving consent in large scale mobile HCI through personalised representations of data. In: Proceedings of the 8th Nordic Conference on Human-Computer Interaction: Fun, Fast, Foundational, pp. 471–480. ACM (2014)

16. Oulasvirta, A., Pihlajamaa, A., et al.: Long-term effects of ubiquitous surveil-lance in the home. In: Proceedings of the 2012 ACM Conference on Ubiquitous Computing, pp. 41–50. ACM (2012)

17. Owen, R., Macnaghten, P., et al.: Responsible research and innovation: From science in society to science for society, with society. Science and Public Policy 39(6), 751–760 (2012)

18. Pagliery, J.: FBI lets suspects go to protect 'Stingray' secrets, CNN, http://money.cnn.com/2015/03/18/technology/security/police-stingray-phone-tracker/ (accessed on March 24, 2015)

19. Schoeman, F.D.: Philosophical dimensions of privacy: An anthology. Cambridge University Press (1984)

20. Tur, G., De Mori, R.: Spoken language understanding: Systems for ex-tracting semantic information from speech. John Wiley & Sons (2011)

21. Waibel, A., Schultz, T., et al.: SMaRT: The smart meeting room task at ISL. In: Proceedings of, IEEE International Conference on Acoustics, Speech, and Signal Processing (ICASSP), vol. 754, pp. IV-752–IV-755. IEEE (2003)

22. Whittaker, S., Hirschberg, J., et al.: SCANMail: a voicemail interface that makes speech browsable, readable and searchable. In: Proceedings of the SIGCHI Conference on Human Factors in Computing Systems, pp. 275–282. ACM (2002)

23. Wyatt, D., Choudhury, T., et al.: Inferring colocation and conversation net-works from privacy-sensitive audio with implications for computational social sci-ence. ACM Transactions on Intelligent Systems and Technology (TIST) 2(1), 7 (2011)

24. Yankelovich, N., Levow, G.-A., et al.: Designing SpeechActs: Issues in speech user interfaces. In: Proceedings of the SIGCHI Conference on Human Factors in Computing Systems, pp. 369–376. ACM Press/Addison-Wesley Publishing Co. (1995)

Internet and Innovation

Communication Forms and Digital Technologies in the Process of Collaborative Writing

Kaja Scheliga[✉]

Alexander von Humboldt Institute for Internet and Society, Berlin, Germany
kaja.scheliga@hiig.de

Abstract. The increasing use of digital technologies for research purposes is affecting scholarly communication. This analysis of a nationwide survey with academics in Germany shows that, in the process of collaborative writing, co-authors communicate across multiple channels. Communication forms based on digital technologies are used complementarily to face-to-face communication. Furthermore, it is evident from the data that the more digital technologies researchers use, the more face-to-face meetings they hold.

Keywords: Digital technologies · Communication · Face-to-face · Collaboration · Internet science

1 Introducing Digital Scholarship

Research, especially in science, technology, engineering and mathematics, is a collaborative effort [1]. Collaboration, however, is not limited to these areas of research; the social sciences, arts and humanities are also affected. There is a growing number of academic papers written in co-authorship [2,3]. At the same time, there is a growing number of digital technologies that are increasingly used for scholarly communication [4,5]. Scholarship is becoming digital and is subject to transformations throughout that process [6,7,8,9]. In light of the increasing use of digital technologies for research purposes, is there a tendency for communication between researchers to be mediated via technologies, consequently leading to a decline of face-to-face meetings? In this paper I explore both digital and direct communication forms between researchers in Germany against the background of collaborative writing of academic papers. I compare the communication forms most frequently used by researchers who collaborate with co-authors working in the same place and those working in different places. I examine whether the ubiquity of digital technologies affects communication between co-authors, leading to communication that is largely mediated via digital technologies and thereby reducing face-to-face contact. The data shows that communication forms based on digital technologies do not lead to a decline of face-to-face communication among co-authors. I argue that researchers use multiple communication channels in the process of collaborative writing. Digitally mediated communication forms are used complementarily to face-to-face meetings. Drawing on my data analysis, I highlight a positive correlation between the use of digitally mediated

© Springer International Publishing Switzerland 2015
T. Tiropanis et al. (Eds.): INSCI 2015, LNCS 9089, pp. 113–122, 2015.
DOI: 10.1007/978-3-319-18609-2_9

communication forms and face-to-face meetings. The more researchers communicate digitally the more they communicate directly.

2   Shifting Towards Digitisation

In light of the ubiquitous presence of digital technologies in the developed world, it can be argued that there is a gradual shift from a historical society that relies on information and communication technologies to record and transmit data, to a hyperhistorical society that depends on information and communication technologies to record, transmit and process data [10]. Furthermore, the transition towards a hyperhistorical state evokes the notion of technology-in-betweenness [11]. Applied to digital scholarship, technology-in-betweenness means that there is a digital technology between collaborating researchers. Consequently, scholarly communication is mediated via digital technologies. This can also be referred to as hyperpersonal communication [12].

3   Science 2.0 Survey

The empirical basis of my analysis is the quantitative Science 2.0 Survey (2014). It was conducted online as part of the Leibniz Research Alliance Science 2.0. The sample includes researchers from various disciplines, from universities as well as from research institutes across Germany. The survey consisted of a main part concerning communication forms and the use of online tools in academia (N=2083) and an optional part concerning online text editors (N=1339). The analysis presented is based on the data from the optional part of the survey and is supplemented by demographic data from the main part of the survey. The statistical analysis was performed using IBM Statistics SPSS 22. The sample is not representative but offers good coverage of the academic landscape in Germany.

4   The Researchers

The respondents to the Science 2.0 survey (N=1339) are researchers in Germany. They cover all age groups, although the majority of the researchers are between 25 and 35 years old (Fig. 1). Correspondingly, a considerable proportion of respondents are at a relatively early stage of their academic career – many are doctoral students or work in research staff and postdoc positions (Fig. 2). However, 14% of the sample consists of professors and junior professors, which thus counteracts the skew towards "digital natives". In terms of academic disciplines the largest fraction of respondents work in the field of mathematics, the natural sciences, and computer science. Nevertheless, the linguistic- and cultural sciences as well as law, economics, the social sciences and engineering are also well represented (Fig. 3). There is a fairly representative distribution with regards to gender: 45% of respondents are female and 55% are male.

Age in years

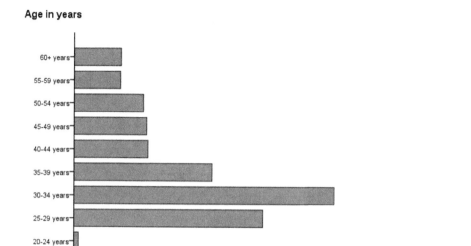

Fig. 1. Age of the researchers

Academic position

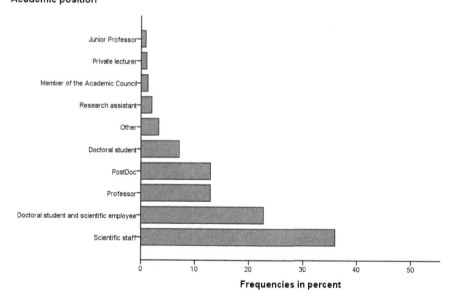

Fig. 2. Academic position of the researchers

Academic discipline

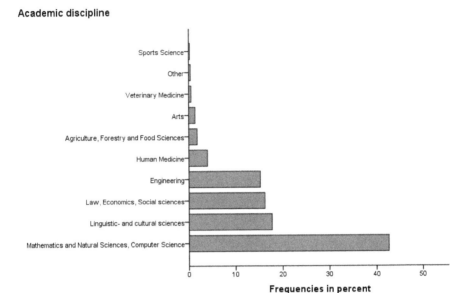

Fig. 3. Academic discipline of the researchers

5   Communication Forms among Researchers

In exploring the impact of digital technologies on forms of communication, postal mail is an interesting case. In the context of collaborative writing, 21% of researchers claim to use postal mail as a form of communication with co-authors who work at a different place. Compared with 91% of email usage, this shows that while mail as a classical form of communication is still being used by researchers, it has been pushed back by the high usage of email.

Another case worth considering is the use of the telephone in comparison with VoIP technologies such as Skype. While on the surface, both technologies transmit the voice of the sender to the receiver, the range of functionalities and the conventions of use are different. VoIP services make it easy to add participants to the conversation and to exchange additional information via the chat channel. Not all researchers, however, are online all the time. Furthermore, even with a stable internet connection the quality of the conversation is unpredictable and subject to disruptions. The telephone, in contrast, is a more reliable communication medium, and it is therefore still frequently used by researchers. In this case there are no signs of VoIP technologies replacing the telephone, both technologies are used in parallel.

Correspondingly, the question is whether there are any indications of digital technologies that imitate direct human contact, such as videoconferencing, reducing face-to-face meetings. The data shows that only 46% of researchers who work at different places use videoconferencing while 92% of researchers have face-to-face meetings. Contrary to the initial assumption that scholarly communication is mediated by digital technologies, this clearly indicates that digital technologies are used as a complementary form of communication.

5.1   Communication Forms in the Same Place

The three most frequently used forms of communication among researchers in the process of collaborative writing are face-to-face meetings, email, and telephone (see Fig. 4). Face-to-face meetings are the communication form most frequently used among co-authors who work at the same place. In total, 93% of researchers have face-to-face meetings during the process of collaborative writing and 64% stated that they have face-to-face meetings very often. Communication via email is used very often by 54% of researchers, and total use of email amounts up to 93%. As mentioned above, the telephone is still an important tool for researchers. In total, it is used by 85% of researchers, while 26% of researchers use it very often.

In contrast to these three established communication forms, newer forms of communication based on digital technologies are used to a smaller extent by researchers working at the same place. Services such as chat or instant messaging are used by 29% of researchers, VoIP by 24% of researchers, and videoconferencing by 16% of researchers. Thus, new digital communication forms are used less frequently than face-to-face meetings, email, and telephone.

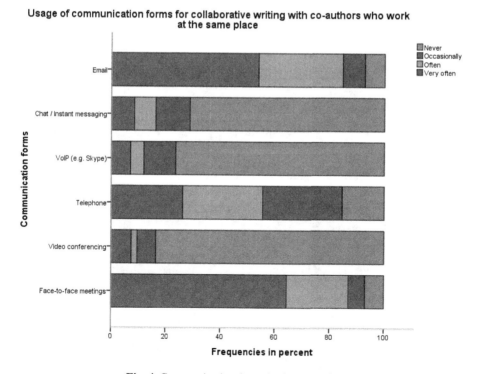

Fig. 4. Communication forms in the same place

5.2   Communication Forms Between Different Places

As is the case for co-authors who work at the same place, co-authors who work at different places also predominantly use email, telephone, and face-to-face meetings to communicate with one another (see Fig. 5). When collaborating at different places, 75% of researchers use email very often. In total, 91% of researchers rely on email when communicating with their co-authors. With physical distance between co-authors, the use of the telephone increases to 92% in total, and 28% of researchers use it very often. In spite of working in different places, face-to-face meetings play an important role. 92% of researchers use this form of communication to collaborate and 22% of researchers hold face-to-face meetings very often.

The usage of newer digital communication forms is higher for co-authors who work in different places than for those who share a workspace. In total, 58% of researchers use VoIP technologies, 46% use videoconferencing, and 37% use chat or instant messaging services.

As the descriptive analysis shows, the general tendencies concerning the usage of communication forms during the process of collaborative writing are the same for co-authors working both at the same and at different places. Established communication forms are used more frequently than those relying on newer digital technologies.

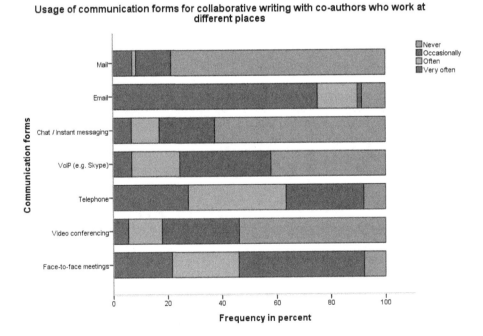

Fig. 5. Communication forms between different places

5.3   The Relation Between Digital and Face-to-Face Communication

In order to thoroughly examine the relation between communication forms based on digital technologies and face-to-face meetings I conducted a statistical analysis. I constructed an index variable, grouping together all variables referring to communication forms based on digital technologies (videoconferencing, VoIP, chat, email, telephone), and correlated it with the variable capturing the frequency of face-to-face meetings. The correlation encompasses communication forms among researchers working both at the same and at different places. The results are summarised in Table 1. Communication forms based on digital technologies and face-to-face meetings positively correlate with each other. This suggests that the more digitally mediated forms of communication researchers use the more they also hold face-to-face meetings. The correlation is stronger between digitally mediated forms of communication and face-to-face meetings for researchers working in different places than for those who are working at the same place. Generally, however, the correlation is relatively weak due to low values of the correlation coefficients. Nevertheless, it is statistically significant.

Table 1. Correlations between digital communication and face-to-face (F2F) communication

Correlations		F2F (@same place)	F2F (@different place)
DigitalCommunication (@same place)	Pearson Correlation	,126**	,255**
	Sig. (2-tailed)	,000	,000
	N	1338	1338
DigitalCommunication (@different place)	Pearson Correlation	,167**	,306**
	Sig. (2-tailed)	,000	,000
	N	1338	1338
**. Correlation is significant at the 0.01 level (2-tailed).			

Correlations (Spearman's rho)		F2F (@same place)	F2F (@different place)
DigitalCommunication (@same place)	Correlation Coefficient	,074**	,187**
	Sig. (2-tailed)	,007	,000
	N	1338	1338
DigitalCommunication (@different place)	Correlation Coefficient	,138**	,142**
	Sig. (2-tailed)	,000	,000
	N	1338	1338
**. Correlation is significant at the 0.01 level (2-tailed).			

The initial question of whether communication between co-authors is largely mediated via digital technologies, consequently leading to a decline of face-to-face meetings, is not supported by the available data. Instead, the data shows that communication forms based on digital technologies are used in a complementary fashion. Researchers communicate using various – both digital and non-digital – channels simultaneously. Moreover, the more digital technologies researchers use the more they tend to meet face-to-face.

6   Discussing Communication Forms

It is clear that digital technologies have an impact on scholarship in general and scholarly communication in particular. The high penetration of various forms of communication technologies in other areas of work and play may lead us to think that also scholarly communication is largely mediated by digital technologies. While there is a tendency for communication among researchers to increasingly rely on digital technologies, the presented results show that researchers use multiple channels for communicating with their co-authors. The analysis reveals no distinct signs that digital communication forms reduce face-to-face communication. This is also reflected by Borgman who explains that '[n]ew technologies did not result in shifting the balance among stakeholders as radically or rapidly as some had hoped, largely because social practices are much more enduring than are technologies.' [13]. The social aspect of face-to-face communication seems to be of importance for researchers. Face-to-face meetings not only provide researchers with a means to discuss pressing issues, but at the same time, they satisfy the basic need for social interaction. After all, mutual inspiration and trust are crucial elements of successful collaboration.

A possible explanation for the lower usage of newer technologies for scholarly communication can be found in a degree of conservatism among researchers when it comes to adopting new technologies [14]. Another reason could be that many digital technologies have not been developed with research in mind and therefore do not necessarily offer the functionalities most needed by researchers.

The fact that researchers tend to communicate across multiple channels can be explained by the nature of the collaborative writing process. It seems that face-to-face meetings are important at the stages of the writing process where complicated problems are discussed or major decisions concerning the text are taken. At these moments, researchers are likely to want the full array of available information from their co-authors. This is something digital technologies imitate but cannot provide without a loss of information. The positive correlation between usage of digital technologies and face-to-face meetings seems to suggest that digital technologies do not suffice when it comes to discussing complex issues among collaborating researchers.

At other stages of the collaborative writing process that involve tasks of a more organisational nature or the fine-tuning of the text, digitally mediated communication forms are likely to be of greater value. Email not only makes coordination fast and cheap but also leaves a record of the facts that is available to all co-authors. As one researcher elucidates in the context of open science, 'when you use those things [online tools], the conversation you have gets recorded, and you can follow it, you can remember the conversations you had with your colleagues' [15]. This record, in contrast to the alterable memory of what was said during a face-to-face meeting, a telephone conversation or a videoconference, is the same for all collaborators. Even though it can be interpreted in different ways it is not subject to change as is the case with memory.

Moreover, the use of multiple communication channels among co-authors can additionally be explained by the synchronous and asynchronous nature of these communication forms. Email or some forms of chat are asynchronous, which is relevant for co-authors who work at different places and possibly even across different timezones. Face-to-face meetings as well as videoconferencing, and telephone conferences are synchronous, which requires researchers to be available at the same time. These differences suggest that the various communication forms are relevant at different stages of the collaborative writing process and thus serve as complementary communication channels.

7   Conclusion

As I have shown based on an analysis of the Survey Science 2.0, researchers use multiple communication channels in the process of collaboratively writing academic papers. Digitally mediated forms of communication are used complementarily to face-to-face communication. The more digital technologies researchers use for communicating in the process of collaborative writing, the more they tend to meet face-to-face. On the one hand, in light of increased usage of digital technologies for research purposes, a stronger digitisation of communication is likely to occur. On the other hand, we should not expect digital technologies to completely replace face-to-face communication in the realm of research collaboration. What is important for the future is not to force more digital communication tools upon researchers but to continue developing digital technology in such a way that is beneficial to scholarly communication.

Acknowledgements. This research has been supported by the Alexander von Humboldt Institute for Internet and Society. Many special thanks to Daniela Pscheida, Claudia Minet, and all others who collaborated on the Science 2.0 Survey 2014. Thanks to the reviewers of the second EINS Conference on Internet Science for their valuable remarks.

References

1. Bozeman, B., Boardman, C.: Research Collaboration and Team Science. A State-of-the-Art Review and Agenda. Springer, Heidelberg (2014)
2. Katz, J.S., Martin, B.R.: What is research collaboration? Research Policy 26, 1–18 (1997)
3. Wuchty, S., Jones, B.F., Uzzi, B.: The Increasing Dominance of Teams in Production of Knowledge. Report. Science 316(5827), 1036–1039 (2007), doi:10.1126/science.1136099.
4. Science 2.0 Survey 2013 Datareport. Pscheida, D., Albrecht, S., Herbst, S., Minet, C., Köhler, T.: Nutzung von Social Media und onlinebasierten Anwendungen in der Wissenschaft. Erste Ergebnisse des Science 2.0-Survey 2013 des Leibniz-Forschungsverbunds Science 2.0 (2013),
http://nbn-resolving.de/urn:nbn:de:bsz:14-qucosa-132962

5. Science 2.0 Survey 2014 Datareport. Pscheida, D., Minet, C., Herbst, S., Albrecht, S., Köhler, T.: Nutzung von Social Media und onlinebasierten Anwendungen in der Wissenschaft – Ergebnisse des Science 2.0-Survey 2014 (2015), `http://d-nb.info/1069096679/34`

6. Borgman, C.L.: Scholarship in the Digital Age. Information, Infrastructure, and the Internet. MIT Press, Cambridge (2007)

7. Weller, M.: The Digital Scholar. How Technology Is Transforming Scholarly Practice. Bloomsbury Publishing (2011)

8. Nentwich, M., König, R.: Cyberscience 2.0. Research in the Age of Digital Social Networks. Campus, Frankfurt-on-Main (2012)

9. Meyer, E.T., Schroeder, R.: Knowledge Machines. Digital Transformations of the Sciences and Humanities. MIT Press, Cambridge (2015)

10. Floridi, L.: Hyperhistory and the Philosophy of Information Policies. Philosophy & Technology 25(2), 129–131 (2012), doi:10.1007/s13347-012-0077-4

11. Floridi, L.: The 4th Revolution. How the Infosphere is Reshaping Human Reality. Oxford University Press, Oxford (2014)

12. Walther, J.B.: Computer-Mediated Communication: Impersonal, Interpresonal, and Hyperpersonal Interaction. Communication Research 23, 3–43 (1996)

13. Borgman, C.L.: Scholarship in the Digital Age. Information, Infrastructure, and the Internet, p. 65. MIT Press, Cambridge (2007)

14. Procter, R., Williams, R., Stewart, J., Poschen, M., Snee, H., Voss, A., Asgari–Targhi, M.: Adoption and use of Web 2.0 in scholarly communications. Philosophical Transactions of the Royal Society 368(1926), 4039–4056 (2010)

15. Scheliga, K.: Open Science Interview with Carolina Ödman-Govender. Zenodo (2014), https://zenodo.org/record/12342?ln=en#.VOsmwvlfaSp

Ethics of Personalized Information Filtering

Ansgar Koene[✉], Elvira Perez, Christopher James Carter,
Ramona Statache, Svenja Adolphs, Claire O'Malley, Tom Rodden,
and Derek McAuley

HORIZON Digital Economy Research
University of Nottingham, Nottingham, UK
ansgar.koene@nottingham.ac.uk,
{first_name,last_name}@nottingham.ac.uk

Abstract. Online search engines, social media, news sites and retailers are all investing heavily in the development of ever more refined information filtering to optimally tune their services to the specific demands of their individual users and customers. In this position paper we examine the privacy consequences of user profile models that are used to achieve this information personalization, the lack of transparency concerning the filtering choices and the ways in which personalized services impact the user experience. Based on these considerations we argue that the Internet research community has a responsibility to increase its efforts to investigate the means and consequences of personalized information filtering.

Keywords: Privacy · Transparency · Behavior manipulation · RRI · Filter bubble

1 Introduction

In the world of on-line information services the dominant business model is one in which no monetary payment is taken from the users. In order to attract maximum user numbers, information businesses therefore find themselves competing primarily based on the perceived quality of their information provision. Since information quantity is usually virtually limitless (which is why quantity is not a viable option for differentiating from competitors), information overload has become one of the main concerns for users. Perceived quality is therefore primarily determined by the ease with which the user can obtain some information that satisfies their current desires. The development of personalized information filtering therefore represents a logical step in the evolution of on-line information services. For many of the most highly success internet service, like Google, Amazon.com, YouTube, Netflix and TripAdvisor, the recommender system is a key element in their success over rival services in the same sector. Some, like Netflix, openly acknowledge this even to the extent of awarding large prizes for anyone that can improve their recommender system.

The simple logic behind the business case for developing such filtering systems however is not sufficient to put to rest the numerous social and ethical concerns that are introduced by the use of these filters. From a Responsible Research and Innovation (RRI) perspective [1], it is necessary for the internet research community to consider the wider implications of such innovations on society.

© Springer International Publishing Switzerland 2015
T. Tiropanis et al. (Eds.): INSCI 2015, LNCS 9089, pp. 123–132, 2015.
DOI: 10.1007/978-3-319-18609-2_10

One of the social concerns about personalized information filtering that has probably attracted the most attention is the fear that optimizing people's information flows to focus on those things they have previously shown an interests/affinity for may cause a feedback loop by which people become isolated from new information due to a self-reinforcing filter bubble [2, 3]. To what extent this can, or does, happen as a consequence of search engine, social media and news feed personalized filtering, is not yet clear. While [4] provided a theoretical analysis showing that, under certain conditions, such a scenario is possible, little experimental work has been done to verify if the 'filter bubble' scenario is taking place. Under some circumstances, it was shown that a personalized Recommender System for music purchases appeared to widen the user's interests [5] rather than narrowing it. In this context it should also be noted that some recommender systems are being specifically designed to promote 'serendipitous discovery' [6].

Unfortunately, most of the research on the impact of personalization and recommender systems has so far focused on their commercial success in increasing sales (e.g. [7]), web impressions (e.g. [8]), and their ability to increase the consumer interest for niche goods (e.g. [9]). As we have argued in our previous position paper [10], this apparent imbalance in research efforts, seemingly focused on a corporate agenda, is exactly the kind of narrative that led to the GMO crop controversy in the EU in the 1990s which dramatically impacted the funding and public support for the Biosciences. In order to avoid such a public backlash against Internet research it is necessary to show that the research community is not solely interested in furthering a corporate agenda, but rather is seriously engaged with identifying and improving the societal impact of Internet research and innovation.

In the remainder of this paper we will focus on a number of other concerns associated with personalized filtering. The main social and ethical concerns we want to draw attention to in this paper are:

1. The privacy intrusion that is unavoidably linked to the tuning of the user behavior profile models;
2. The lack of transparency concerning the data that is used, how it is gathered and the way the algorithms work;
3. The risks of covert manipulation of user behavior.

2   Brief Review of Recommender Systems

Recommender systems emerged as an independent research area in the mid-1990s. These first recommender systems [11] applied collaborative-filtering which works on the principle that a user who has in the past agreed with certain other users (i.e. given similar ratings, or 'clicked' on similar items) will have similar interests to them and will therefore find relevant and recommendations for items that these users rated highly. Modern recommender systems using (combinations of) various types of knowledge and data about users, the available items, and previous transactions stored in customized databases. The knowledge and data about the users is collected either through explicitly ratings by the users for products, or are inferred by interpreting user actions, such as the navigation to a particular product page which is interpreted as an implicit sign of preference for the items shown on that page.

The two main classes of recommender systems are:

- Content-based, where the system learns to recommend items that are similar to the ones that the user liked in the past. The similarity of items is calculated based on the features associated with the compared items. Figure 1 gives a high-level overview of the components and data flow in a content based recommender system.
- Collaborative-filtering, users are given recommendations for items that other users with similar tastes liked in the past. The similarity in taste of two users is calculated based on the similarity in the rating history of the users.
- Community-based, where the system recommends items based on the preference the user's friends. This is similar to Collaborative filtering except that the selection of peers to be used for identifying the recommendation is based on an explicit 'friendship' link instead of being deduced from patterns of similar past behavior. Such 'social recommender' systems are poplar in social-network sites [12].

In practice many of the recommender systems are hybrid systems that try balance the advantages and disadvantages of each class [13]. Collaborative and community based systems, for instance, suffer from an inability to recommend items that have not yet been rated by any of the potential peers of the user. This limitation however does not affect content-based system as long as the new item is supplied with a description of its features, allowing it to be compared to other items that the user has interacted with in the past.

A comprehensive introduction to recommender systems is provided in [14].

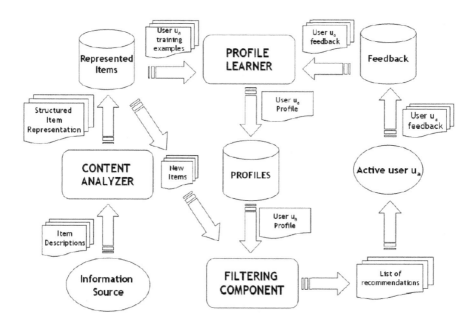

Fig. 1. High level architecture of a Content-based recommender system

3   Privacy Intrusion

The filter parameters that determine the personalized selection and ranking of information constitute an implicit user profile model, which is usually based for a large part on data about the past search and browsing behavior of the users when they previously interacted with the service [15]. To further refine the user profiles, services may also to gather information about the user behavior on other websites through the use of 'tracking cookies' [16] or by purchasing third-party access to such data from other services. Additionally, recommender systems may also use data concerning the behavior of people within the social network of the users [17].

From a privacy and digital human rights perspective, each of these data gathering methods is ethically troubling since they are all surreptitious, to varying degrees. The use of 'tracking cookies' is clearly the most troubling in this respect, however even the logging of users' behavior when they are actively engaging with the information service itself lacks proper informed consent. At best, users may have read something about data logging in the terms-and-conditions they had to agree to when they first signed up to the service. Unfortunately the current reality of Internet usage is that terms-and-conditions policies of Internet sites are rarely read and are generally formulated in ways that are too vague and incomprehensible to constitute a means of gaining true informed consent [18]. Furthermore, it is not realistic to expect users to remain vigilantly aware of the information about data tracking in the terms-and-conditions weeks, months and years after they signed up to the site. The "EU Cookie Law" [19] has gone some way towards providing a frequent reminder of data tracking by websites, however the standard notification of the type:

"By continuing to use this site you consent to the use of cookies on your device as described in our cookie policy unless you have disabled them. You can change your cookie settings at any time but parts of our site will not function correctly without them." [FT.com]

is generally too vague for people to understand, resulting in the same dismissive 'click to close' behavior that people have become accustomed to from the cryptic error/warning messages that are typically generated by software and the terms-and-conditions agreements they did not read.

Beyond the data collection process, the user profile models that form the basis of personalized information filtering pose an additional privacy concern in themselves. The profile models of users are in essence an operationalization of the data mining efforts by the service provider, built to anticipate the user's behavior, interests and desires. Access to a perfect behavior model of a user would in principle enable anyone to predict the user's actions/decisions for a wide range of choices/conditions. Beyond the immediate commercial potential for guiding 'relevant' advertisements to a person, such person profiles could be used to plan targeted phishing campaigns or hacking related social engineering. To find a user's weaknesses it would suffice to query the user's behavior model with a range of choices and observe the predicted responses.

Based on the analysis above, we observe that models of user behavior profiles are necessary for the functioning of personalized recommender systems and that these models are unavoidably linked to a certain level of privacy intrusion. We therefore propose that, within the RRI framework, the Internet research community should focus not only on ways to further fine-tune the recommendations from such systems, but also on developing recommender systems architectures where the user profile model, and the corresponding privacy sensitive information, stays within the access-control domain of the user. One possible method for this might be based on a two-layer architecture where the first layer, hosted by the information service provider, generates non-personalized search/recommendation results, which are provided to a second layer, hosted locally on the user's device, which ranks the results based on the personalized user profile.

4   Transparency

Due to the commercial advantage which information service providers hope to achieve through the use of personalized information filtering, the information about how exactly their filtering is done is not made publicly available. This lack of transparency however also makes it impossible for users to gain a full understanding of how their data is being gathered and used, thus preventing them from truly providing informed consent. A key concern in this regard is the fact that most of the information service businesses do not earn their money from the users but rather from corporate customers who pay access to user data in order to push targeted advertising. There is thus a valid argument for demanding greater oversight into functioning of the information filtering in order to guarantee that the information provided to users best servers their needs. Due to the personal nature of the data used for the behavior profiles, and the user models themselves, there also needs to be transparent verification that these are store and handled in accordance with the jurisdictionally appropriate privacy-related regulation and laws (e.g. [20][21]). While this might be viewed as primarily the responsibility of regulatory oversight authorities, we propose that this should also be seen as a challenge to the Internet research community to develop tools with which the filtering criteria can be probed without access to the underlying code/algorithm. An obvious approach for developing such tools might be to follow existing Black-box testing practices which are commonly used in software and general systems development for evaluation and functional testing [22]. An example for this is provided by [23], where a Black-box testing approach was used to investigate what kind of recommendation schemes were exploited by various movie recommendation systems.

In accordance with the principle of public engagement and dialog concerning research and innovation, the RRI framework also suggests that the Internet research community should make user-friendly versions of recommender system testing kits available to the general public to enable people to evaluate for themselves if they find the level of profile personalization used by the recommender system acceptable of not.

5   Behavior Manipulation

Directly relating to the concerns about lack of transparency, as well as the issue of 'filter bubbles' is the question of how much and for which agenda user behavior is being manipulated by the use of personalized filtering. To a certain extent, behavior manipulation is unavoidable in any information presentation system since people will invariably select the first items on a list more often than those much further down. Since it is impossible to place all information at the top of the list, the act of ranking involves a behavior manipulation. Provided the manipulation is based on mutual consent, there is nothing wrong with this. It is in fact the desired function of a search or recommendation system, as long as the user knows and agrees to the ranking criteria that are used by the algorithm. Advertising, of course is all about attempting to persuade, i.e. manipulate, potential consumers into purchasing the product/service of the advertising agency's client. The dominant business model of advertising funded online information service therefore constitutes a significant conflict of interests at the heart of the filter criterion selection.

As long as the information filtering, and advertising targeting, were based on global statistical criteria it was usually relatively easy for users to judge if the information they were provided with was advertising motivated. The general coarseness of the match between the provided and the desired information also meant users engaged a more critical attitude towards evaluating the search results. The introduction of personalized information filtering however is improving the personalized targeting success of advertisements at least as fast as the general filtering success. Also, due to the generally improved information services, people are less critical in their final selection.

In case there was any doubt, the willingness of information service providers to engage in manipulating their information filtering for purposes other than the service to the user was clearly demonstrated by the "Facebook news feed experiment" [24].

Once again, there is undoubtedly a role for regulatory oversight concerning these conflicts of interest, similar to such regulation in other media. The fact that the personalized filtering and advert targeting systems are developed by Internet researchers, however, means that the Internet research community will undoubtedly be implicated in any future scandals about manipulation of personalized information filtering, as it already was with [24]. In order to mitigate the impact of such events it is therefore important for the Internet research community to be visibly engaged with RRI agenda.

The challenges in this case are simultaneously daunting and yet very familiar: how to prove beyond reasonable doubt that a statement/recommendation is objectively true and unbiased. Following the example of legal court cases, the first step might be to ask the defendant, i.e. the service provider who controls the recommender system, to provide evidence concerning the basis for the recommendation.

In a bid to gain people's trust, and interest in the offered recommendations, various recommender systems, e.g. Amazon.com, already provide some level of evidence by informing the users why certain recommendations are given with statements like: "Customers Who Bought This Item Also Bought". While such information can clearly help users to better understand, and thus evaluate and trust, recommendations it does not fully address concerns about possible behavior manipulation. Such concerns can only be addressed through the provision of trusted-third-party involvement, either in the form of regulatory oversight or by providing tools which can allow users to test the recommender system for un-accounted for recommendation biases.

6   Evidence of Public Concern about Recommender Systems

In this section we summaries a number of news stories that illustrate the level of concern, rightly or wrongly, over lack of transparency and potential bias in recommender systems.

In 2011, the US Federal Trade Commission started an investigation into possible search results bias by Google. It took two years of investigating before Google was cleared of the charges [25].

In February 2014, Google agreed to a settlement with European competition regulators following years of legal struggles with antitrust authorities, starting in 2010, concerning complaints that Google search rankings unfairly favored Google products [26].

In 2010, Netflix decided to cancel the Netflix Prize sequel after the US Federal trade Commission raised concerns about Netflix user privacy and a lawsuit was filed against Netflix [27]. The Netflix Prize competition, and its planned sequel, challenged competitors to develop improved recommendation algorithms based on a published set of anonymized Netflix user data of the type. One of the reasons for the privacy concerns was the publications in 2008 of a paper showing that the data supplied for the recommender algorithms by the Netflix prize dataset was rich enough to allow it to be de-anonymized [28].

7   Conclusion

Personalized information filtering by online search engines, social media, news sites and retailers represents a natural evolution in the development towards ever more finely tuned interaction with the users. Even leaving aside concerns about individual and social consequences of possible 'filter bubbles', the user profiling required to achieve this personalization raises numerous ethical issues around privacy and data protection. Further concerns arise due to the lack of transparency and the potential for increasingly covert manipulation of user behaviour in favour of the commercial interests of the predominantly advertising based business models of information services. Due to the frequently close involvement of the large information service providers with the Internet research community, there is a growing risk that scandals related to personalized information filtering by corporations might triggering a controversy and public backlash similar to the one that hit GM crops in Europe in the 1990s. In order to avoid such a controversy it is essential to retain the confidence and trust of the public by actively engaging with the Responsible Research and Innovation agenda and pro-actively working to mitigate these issues. In order to achieve this we propose a research programme aimed at:

- identifying and studying the socio-psychological impact of personalized filtering;
- helping people to understand and regulate the level of privacy intrusion they are willing to accept for personalized information filtering;
- developing a methodology to probe the subjective 'validity' of the information that is provided to users based on their own interests;
- engaging with corporate information service providers to reinforce ethical practices.

Project elements for such a research programme might include

- Technical development of tools:
  - Black-box testing kit for probing the characteristics of the user behavior profiles used in recommender systems.
  - Recommendation bias detection system for identifying user behavior manipulation
  - A two-layer recommender architecture that de-couples the delivery of non-personalized information by service providers from a user owned/controlled system for personalized ranking of the information.

- Psycho-social research on the impact of personalized information filtering on:
  - General exploration-exploitation trade-off in action selection
  - Attitudes towards trust and critical evaluation of information

- Cybersecurity:
  - Protection against mal-use of personalized recommender systems for phishing related social engineering

- Policy:
  - Development of guidelines for responsible innovation and use of recommender systems, protecting the privacy and freedom of access to information of users.

- Public engagement:
  - Develop educational material to help people understand how recommendations they receive from search engines, and other recommender systems, are filtered so that they can better evaluate the information they receive.

Acknowledgement. This work forms part of the CaSMa project supported by ESRC grant ES/M00161X/1. For more information about the CaSMa project, see http://casma.wp.horizon.ac.uk/ .

References

1. Sutcliffe, H.: A report on Responsible Research & Innovation. Matter (2011), http://ec.europa.eu
2. Pariser, E.: The Filter Bubble. Penguin Books (2011)
3. Sunstein, C.R.: Republic.com. Princeton University Press, Princeton (2007)

4. Van Alstyne, M., Brynjolfsson, E.: Global village or cyber-Balkans? Modeling and measuring the integration of electronic communities. Management Science 51(6), 851–868 (2005)

5. Hosanagar, K., Fleder, D., Lee, D., Buja, A.: Will the Global Village Fracture into Tribes? Recommender Systems and their Effects on Consumer Fragmentation. Management Science 60, 805–823 (2014)

6. Cao Zhang, Y., Ó Séaghdha, D., Quercia, D., Jambor, T.: Auralist: Introducing Serendipity into Music Recommendation. In: Proc. of the 5th ACM Int. Conf. on Web Search and Data Mining (WSDM 2012), Seattle, Washington, USA, February 8-12 (2012)

7. De, P., Hu, Y.J., Rahman, M.S.: Technology usage and online sales: an empirical study. Management Science 56(11), 1930–1945 (2010)

8. Das, A., Datar, M., Garg, A., Rajarm, S.: Google news personalization: scalable online collaborative filtering. In: Proc. of the 16th Int'l World Wide Web Conference, pp. 271–280 (2007)

9. Fleder, D., Hosanagar, K.: Blockbuster culture's next rise or fall: the impact of recommender systems on sales diversity. Management Science 55(5), 697–712 (2009)

10. Koene, A., Perez, E., Carter, C.J., Statache, R., Adolphs, S., O'Malley, C., Rodden, T., McAuley, D.: Research Ethics and Public Trust, Preconditions for Continued Growth of Internet Mediated Research. In: 1st International Conference on Information System Security and Privacy (ICISSP), February 9-11 (2015)

11. Goldberg, D., Nichols, D., Oki, B.M., Terry, D.: Using collaborative filtering to weave information tapestry. Commun. ACM 35(12), 61–70 (1992)

12. Golbeck, J.: Generating predictive movie recommendations from trust in social networks. In: Stølen, K., Winsborough, W.H., Martinelli, F., Massacci, F. (eds.) iTrust 2006. LNCS, vol. 3986, pp. 93–104. Springer, Heidelberg (2006)

13. Burke, R.: Hybrid web recommender systems. In: Brusilovsky, P., Kobsa, A., Nejdl, W. (eds.) Adaptive Web 2007. LNCS, vol. 4321, pp. 377–408. Springer, Heidelberg (2007)

14. Rokach, L., Shapira, B., Kantor, P.B.: Recommender systems handbook, vol. 1. Springer, New York (2011)

15. Speretta, M., Gauch, S.: Personalized search based on user search histories. In: Proceedings of the 2005 IEEE/WIC/ACM International Conference on Web Intelligence, September 19-22, pp. 622–628 (2005), doi:10.1109/WI.2005.114

16. Rohle, T.: Desperately seeking the consumer: Personalized search engines and the commercial exploitation of user data. First Monday, [S.l.] (September 2007), http://journals.uic.edu/ojs/index.php/fm/article/view/2008/1883 ISSN 13960466

17. Ma, H., Zhou, D., Liu, C., Lyu, M.R., King, I.: Recommender systems with social regularization. In: WSDM 2011 Proceedings of the Fourth ACM International Conference on Web Search and Data Mining, pp. 287–296 (2011), doi:10.1145/1935826.1935877

18. Luger, E.: Consent for all: Revealing the hidden complexity of terms and conditions. In: Proceedings of the SIGCHI Conference on Human Factors in Computing Systems, pp. 2687–2696 (2013)

19. The Privacy and Electronic Communications (EC Directive) Regulations (2003), http://www.legislation.gov.uk/uksi/2003/2426/contents/made

20. European Union (EU) Data Protection Directive of (Directive 95/46/EC) (1995), http://eur-lex.europa.eu/legal-content/EN/TXT/HTML/?uri=CELEX:31995L0046

21. OECD Guidelines on the Protection of Privacy and Transborder Flow of Personal Data (C(80)58/FINAL, as amended on 11 July 2013 by C(2013)79), http://www.oecd.org/sti/ieconomy/privacy.htm

22. Beizer, B.: Black-box testing: techniques for functional testing of software and systems. John Wiley & Sons, Inc. (1995)
23. Lee, N., Jung, J.J., Selamat, A., Hwang, D.: Black-box testing of practical movie recommendation systems: A comparative study. Computer Science and Information Systems 11(1), 241–249 (2014)
24. Kramer, A.D.I., Guillory, J.E., Hancock, J.T.: Experimental evidence of massive-scale emotional contagion through social networks. PNAS 111(24), 8788–8790 (2014)
25. Arthur, C.: Google cleared of search results bias after two-year US investigation. The Guardian (January 4, 2013)
26. Miller, C.C., Scott, M.: Google Settles Its European Antitrust case; Critics Remain. The New York Times (February 5, 2014)
27. Netflix official blog announcement (March 12, 2010), http://blog.netflix.com/2010/03/this-is-neil-hunt-chief-product-officer.html
28. Narayanan, A., Shmatikov, V.: Robust de-anonymization of large datasets (how to break anonymity of the Netflix prize dataset). University of Texas at Austin (2008)

IAA: Incentive-Based Anonymous Authentication Scheme in Smart Grids

Zhiyuan Sui$^{(\boxtimes)}$, Ammar Alyousef, and Hermann de Meer

University of Passau, Innstr. 43, 94032 Passau, Germany
{suizhiyu,ammar.alyousef,demeer}@fim.uni-passau.de

Abstract. The traditional energy consumption calculation heavily relies on manual work, which is inefficient and error-prone. The Smart Grid, which integrates information and communication technologies into the electrical grid to gather information and manage energy production and consumption, may be a solution to this challenge. However, the resulting complex infrastructure and profusion of information may open up new attack vectors exploitable by malicious parties that could attack the grid itself or violate its consumers' privacy. In this paper, we indicate the increasing interests in providing conditionally anonymous authentication in the Smart Grid systems. While the consumption report stays anonymous, the consumers who voluntarily curtail their energy consumption, can confirm their curtailments in the scheme. Moreover, compared with the existing conditionally anonymous authentication schemes, our scheme is more efficient in computational and communication overhead for Smart Grid systems.

Keywords: Smart Grids · Anonymous authentication · Demand and response · Privacy preservation · Incentive

1 Introduction

Smart Grid systems combine advanced communication and automated control technologies in order to increase flexibility and resilience of the infrastructure, save energy and reduce CO_2 emissions [1]. The integration of new technologies however leads to a completely new infrastructure, where the formerly isolated electrical grid, which is currently one of the most critical infrastructures, is blended with methods from Information and Communication Technologies (ICT). Households are equipped with intelligent smart meters and smart appliances and also the energy provider enhances its systems with new hardware and IP-based networking [2]. Smart meters measure energy consumption in a much higher temporal resolution than conventional meters and send the gathered energy consumption data to the utility provider in order to achieve better

This work has received support by the EINS (FP7 NoE, grant no. 288021). The research leading to these results was supported by the "Bavarian State Ministry of Education, Science and the Arts" as part of the FORSEC research association. The first author is supported by the Chinese Scholarship Council.

© Springer International Publishing Switzerland 2015
T. Tiropanis et al. (Eds.): INSCI 2015, LNCS 9089, pp. 133–144, 2015.
DOI: 10.1007/978-3-319-18609-2_11

monitoring, control and stability of the Smart Grid. At the power shortage time, the utility provider provides incentive payments to consumers for reducing their loads during reliability triggered events, but curtailment is voluntary. This new combination of energy network and ICT technology puts the security of the Smart Grid in question as it creates new ways to attack and tamper with the highly critical energy supply [3]. Two of the most challenging tasks are privacy and security. From an end user's point of view, the fine-granular energy consumption readings of a smart meter could be used to spy on and expose an user's activities at home. As shown in [4], the Smart Grids, which gather and analyze such information, lead to the large-scale creation of user profiles without a victim's consent or even his knowledge. This in turn could lead to personalized advertisements or discrimination against a user who is negatively classified according to his energy usage behavior. Therefore, the protection of a user's privacy is an essential necessity in the Smart Grid to achieve an adequate overall acceptance of this technology. On the other side, the security on the demand-response communication also needs to be ensured. As consumption data are transmitted through networks, the number of attack vectors vastly increased with the introduction of networked ICT in electrical meters [5].

To find a technological approach that provides both privacy as well as security was a great research interest over the last couple of years. While there were many different approaches in this direction [6] [7], the reward distribution relies on the trusted third party so far. Once it is attacked, consumers' privacy will be leaked. In this paper, we design an anonymous authentication scheme for incentive-based demand response programs, named IAA. Specifically, the contributions of IAA are twofold.

1. Firstly, IAA can achieve strong anonymity and reward support. The electricity utility broadcasts the energy usage instruction to consumers and advises them to reduce their energy consumption by an acceptable percentage, when it finds an imbalance between the energy consumption and production. The willing consumers will revoke their anonymity and get their corresponding rewards, while no other party is able to identify the source of other consumers' usage data.
2. Secondly, compared with previous anonymous authentication schemes, which can provide similar security properties, for one thing, IAA is identity-based; for another, the computational and communication overhead is independent with the number of consumers in IAA. Therefore, it is more suitable for large group Smart Grid systems.

The remainder of this paper is structured as follows: Section 2 describes the works that employ cryptosystem to achieve the security in Smart Grids up till now. In Section 3, the preliminaries, which are later on required in this paper, are explained in detail, while Section 4 explains our proposed scheme that features both anonymity and security. The security requirements are proved in Section 5. Section 6 compares the computational and communicational performance of our scheme with previous works. Section 7 concludes this paper.

2   Related Work

In order to achieve security in the Smart Grid systems, identity based signature schemes and anonymous authentication schemes are widely utilized.

Identity based signature (IBS) was introduced by Shamir [8]. The public key is generated from the user's identity in an IBS. IBS eliminates the overhead for checking the validity of the certificates. In reference [9], So et al. propose an IBS for Smart Grids, which does not require pre-device software setup from the users, and simplifies the key management mechanism. Nicanfar et al. [10] propose an efficient authentication and key management mechanism for Smart Grid communication. It prevents from various attacks while reducing the management overhead. Li et al. [11] integrate a homomorphic encryption algorithm and IBS to ensure the privacy and trustworthiness in Smart Grids. However, the key pairs are generated from a key generation server (KGS) in IBS. It assumes that the KGS is completely trustworthy.

Anonymous authentication schemes, e.g. group signatures and ring signatures, are also widely used in Smart Grids for privacy and security. In [12], He et al. employ the group signature to distribute the trustworthiness for the Smart Grid. Only the law authority can ask the information from electricity utility and group manager to revoke the anonymity of the target users. However, it assumes that the law authority is fully trustworthy. Chu et al. [13] construct an anonymous authentication to inquire the usage history records. This scheme cannot ensure the voluntary consumers, who curtail their consumption, can get their rewards. More than that, the computational cost and communication overhead are increasing with the number of the members in the ring signature. There are usually hundreds of smart meters in the Smart Grid system, while the computational resources of smart meter are limited.

In this paper, in conjunction with IBS, we construct an incentive-based anonymous authentication scheme to ensure the demand-response communication between the electricity utility and smart meters. Consumers are categorized according to their behavior. The consumers, who follow the electricity utility's instructions, can get their rewards, while others still stay anonymous. Compared with the previous schemes, on the one hand, IAA is third party free; on the other hand, it is more efficient in terms of the communicational and computational cost.

3   Preliminaries

We list several necessary notations and definitions for our work in this section.

3.1   Bilinear Map

In IAA, we employ the bilinear map to construct an anonymous authentication scheme. The bilinear map operation is based on elliptic curves. κ is a random integer. Input κ, a prime number p of size κ, is selected. $\mathbb{G}$ is a cyclic additive groups of order p. $\mathbb{G}_T$ is a multiplicative group of order p. P is a generator of

$\mathbb{G}$. A function $e : \mathbb{G} \times \mathbb{G} \to \mathbb{G}_T$ is said to be a bilinear map if it satisfies the following properties:

1. **Bilinearity:** $e(aP, bP) = e(P, P)^{ab}$ for all $P \in \mathbb{G}$ and $a, b \in \mathbb{Z}_p^*$.
2. **Non-degeneracy:** $e(P, P) \neq 1$.
3. **Computability:** $e(P, P)$ is efficiently computable, for all $P \in \mathbb{G}$.

3.2   Computational Assumptions

IAA is based on three computational assumptions

1. **Gap-Discrete Logarithm (Gap-DL) Assumption.** There is no probabilistic polynomial time (PPT) algorithm that can compute a number $x \in \mathbb{Z}_p^*$ from a tuple (T, μ), where, $\mu \leftarrow \mathbb{G}$ and $T = x\mu$.
2. **Decisional Diffie-Hellman (DDH) Assumption.** There is no PPT algorithm that can distinguish between a tuple $(\mu, x\mu, \mu', T)$ and a tuple $(\mu, x\mu, \mu', x\mu')$, where $\mu, \mu', T \leftarrow \mathbb{G}$ and $x \leftarrow \mathbb{Z}_p^*$.
3. $q-$**Strong Diffie-Hellman ($q-$SDH) Assumption.** There is no PPT algorithm that can compute a pair $(c, (1/(x + c))P)$, where $c \in \mathbb{Z}_p^*$, from a tuple $(P, xP, ..., x^q P)$, where $P \leftarrow \mathbb{G}$ and $x \leftarrow \mathbb{Z}_p^*$.

3.3   Zero Knowledge Proof

IAA extensively employs non-interactive zero knowledge proof (ZKP) protocol. ZKP is first proposed by Goldwasser et al. [14]. The purpose of ZKP, denoted as $PK\{(x) : C = xP\}$, is to help a prover convince a verifier that he holds the knowledge x, without leaking any information about x during the verification process. ZKP are widely utilized in digital authentication schemes, e. g. Schnorr Signature [15].

3.4   BBS+ Signature

BBS+ signature is initiated by Au et al. [16]. BBS+ signature is proved unforgeable without random oracles under $q-$ SDH assumption. It allows generation of a single signature for a message. Nguyen constructs an efficient knowledge proof of the signature and message without revealing any useful information about either [17].

3.5   Network Model

In our network model, we assume that the usage data is transmitted by the wide area network (WAN). The network model mainly consists of two entities: the electricity utility (EU) and the smart meter (SM). The communication between SMs and the EU is through wireless network technology. We assume that each EU communicates with multiple SMs in a concrete area, and the number of SMs is large enough for each SM to cloak its real identity.

The SM is the energy consumption reporting device present at each consumer's site. The SM reports consumers' energy usage report with the transformed credentials to the EU regularly. Therefore, no one can link the usage report to its source. In IAA, the cooperative consumer would like to curtail his consumption and prove his cooperation. Because each SM corresponds to a concrete consumer, we assume that the cooperative consumer can be confirmed via the real identity of the SM.

The EU is an infrastructure that is controlled by the electricity company and is in charge of the SMs in a concrete area. It collects and analyzes the usage data from SMs periodically, and broadcasts consumption related instructions to customers, according to the usage data. It is unnecessary for the EU to cloak its real identity. In our scheme, the EU's real identity is always considered public.

3.6 Security Requirements

In our anonymous authentication scheme, the main aim is to ensure trustworthiness of the data from both EU and SMs while ensuring the privacy of legitimate users habits. IAA can satisfy the following security requirements simultaneously:

1. The adversary is able to modify neither the consumption reports from SMs, nor the instruction from the EU (data integrity).
2. The EU can determine whether the signature derives from a legitimate source (identity authentication).
3. The consumer, who does not follow the instructions, cannot produce a valid signature to cheat out of rewards (reward-support).
4. The adversary cannot trace an uncooperative consumer's identity using the usage report (strong anonymity).

4 Proposed Scheme

IAA consists of the following procedures: setup algorithm, joining, anonymous report, demand generation and voluntary response protocols. In the setup algorithm, the EU generates its key pair and publishes its public key. During the joining procedure, each SM cloaks its secret key in the credential with Gap-DL assumption. And then, the EU authorizes the credential with BBS+ signature. Finally, the SM obtains a key pair authorized by the EU. After joining into the Smart Grid system, the SM reports its energy consumption data regularly (normally every 15 minutes). The SM transforms all its credentials, and proves its secret information to the EU by zero knowledge proof. Therefore, the EU can confirm whether the signature is from a legitimate SM without the SM's identity. The EU broadcasts the instructions with the signature to the SMs, once it finds that the energy consumption is too large to produce in the demand generation protocol. The consumer checks the timestamp and confirms that the signature is valid. During the voluntary response protocol, if the consumer would not like to cooperate, he just ignores the instruction, and his usage profile is still under anonymity; otherwise, he curtails the energy consumption and proves his curtailment with IBS during voluntary response.

4.1 Setup

The EU executes the setup algorithm to generate its long term key pair:

1. On input κ, the bilinear pairing instance generator returns a tuple $(p, \mathbb{G}, \mathbb{G}_T, e, P)$ as defined in Subsection 3.1.
2. Randomly choose three elements $Q, H, G \leftarrow \mathbb{G}$ and an integer $\gamma \leftarrow \mathbb{Z}_p^*$, hide its secret key in P_{pub}: $P_{pub} = \gamma P$.
3. Choose collision resistant hash functions $\mathcal{H}_1 \colon \{0, 1\}^* \rightarrow \mathbb{G}$; $\mathcal{H}_2 : \{0, 1\}^* \rightarrow \mathbb{Z}_p^*$.
4. Keep its secret key γ and publish its public key (P, Q, H, G, P_{pub}) and hash functions $(\mathcal{H}_1, \mathcal{H}_2)$.

4.2 Joining

The joining protocol is carried out between the EU and each SM. Each SM is equipped with a tamper-resistant black box [18]. Each black box has its key pair **(SK, PK)**. The EU has access to the public key **PK**. In additional, each black box would generate an internal private seed specific to itself. The seed is stored securely within the black box and is never disclosed or changed, as the black box is assumed to be tamper-resistant. Additionally, a secure public key signature scheme, including a signing algorithm **sig** and a verification algorithm **ver**, has been selected for a SM with key pair **(SK, PK)**. Each SM shows its real identity and produces its key pair during the following protocol: At first, the SM randomly generates an integer $x \leftarrow \mathbb{Z}_p^*$ as its secret key using its internal seed. Then, the SM computes a commitment C on the value x: $C = xP$ and generates a signature $\sigma =$**sig**$(C\|\text{ID})$. The SM sends $C\|\text{ID}$ as well as its signature σ to the EU. The commitment C essentially binds the SM's secret key x. Upon receiving C, the EU executes the verification algorithm to check the validity of the signature using **PK**. If **ver**$(C\|\text{ID}, \sigma, \textbf{PK})=$valid, the EU computes the credential $\alpha = \mathcal{H}_2(ID)$, $S = \frac{1}{\gamma+\alpha}(C + Q)$ and sends S to the SM. The SM confirms the correctness of the credential by checking equation $e(S, \alpha P + P_{pub}) = e(C + Q, P)$ holds. The SM's secret key is x, and its public key is (C, S).

4.3 Anonymous Report

In order to achieve the almost real-time usage report, a SM and the EU can run the anonymous report protocol to produce a legitimate signature as following: Firstly, by using the knowledge of x, the SM binds the usage data m and the timestamp t with the element T. The SM computes $\mu = \mathcal{H}_1(m\|P\|P_{pub}\|G\|H\|Q\|t)$ and $T = x\mu$. The SM then proves $e(S, \alpha P + P_{pub}) = e(xP + Q, P)$ and $T = x\mu$ with the following non-interactive zero knowledge proof Equation 1:

$$\text{PK}\left\{ \begin{pmatrix} S \\ x \\ \alpha \end{pmatrix} : \begin{matrix} e(S, \alpha P + P_{pub}) = e(xP + Q, P) \\ T = x\mu \end{matrix} \right\} \tag{1}$$

The procedure of the proof is formally described below:

1. The SM randomly picks integers $r, k_0, k_1, k_2, k_3 \leftarrow \mathbb{Z}_p^*$.
2. In order to cloak its identity information, the SM transforms its original credential S into a temporary one $U = S + rH$, where $r \in \mathbb{Z}_p^*$, and calculates $R = rG$, $M_1 = k_1 G$, $M_2 = k_2 G - k_3 R$, $N = k_0 \mu$, $V = e(P, P)^{k_0} e(H, P_{pub})^{k_1} e(H, P)^{k_2} e(U, P)^{-k_3}$.
3. The SM calculates $g = \mathcal{H}_2(T\|R\|U\|M_1\|M_2\|N\|V\| m\|t)$, $s_0 = k_0 + gx$, $s_1 = k_1 + gr$, $s_2 = k_2 + gr\alpha$, $s_3 = k_3 + g\alpha$.

The SM can show that both the temporary credential and the element T correspond to the same key pair x, α and S without leaking any information of them. Given two signatures, it is impossible to determine whether they are produced by the same SM, or to identify the SM. Consequently, anonymity is achieved. In the end, the SM outputs $(T, R, U, g, s_0, s_1, s_2, s_3)$ as the signature.

After the receipt of the usage report, the EU checks the validity of the timestamp. Then, the EU executes the report reading algorithm to check whether the signature does prove the knowledge of a discrete logarithm x as well as the knowledge of the valid credential S.

The EU computes the hash values $\mu = \mathcal{H}_1(m\|P\|P_{pub}\|G\|H\|Q\|t)$ and $M_1' = s_1 G - gR, M_2' = s_2 G - s_3 R, N' = s_0 \mu - gT, V' = e(P, P)^{s_0} e(H, P_{pub})^{s_1} e(U, P_{pub})^{-g} e(Q, P)^g e(H, P)^{s_2} e(U, P)^{-s_3}$, then confirms that equation $g = \mathcal{H}_2(T\|R\|U\|M_1'\|M_2'\|N'\|V'\|m\|t)$ holds. If it holds, the EU accepts the usage report; otherwise, the EU rejects the usage report.

4.4 Demand Generation

Once the EU finds that the energy consumption is larger than production, it executes the instruction generation protocol to advise some consumers to shut down their appliances:

The EU first defines the instruction (λ, t_n). The EU then employ Schnorr Signature to generate a valid signature to prove its identity: It randomly picks $k_4 \leftarrow \mathbb{Z}_p^*$, computes $W = k_4 P$, $f = \mathcal{H}_2(\lambda\|t_n\|W\|P\|P_{pub}\|t)$ and $s_4 = k_4 - f\gamma$. At last, the EU broadcasts the instructions and the signature $(\lambda, t_n, s_4, f, t)$ to all SMs.

Upon receiving the usage instructions, the SM checks whether the timestamp and the instruction are valid. It computes $W' = fP_{pub} + s_4 P$, checks whether $f = \mathcal{H}_2(\lambda\|t_n\|W'\|P\|P_{pub}\|t)$. If they hold, the SM informs the consumer to shut down his appliances; otherwise, it just rejects the instruction and signature.

4.5 Voluntary Response

After receiving the instruction, if the consumers would like to curtail their consumption by λ, they will execute the voluntary response protocol with the EU.

The SM should confirm that its current usage data m and the usage data m^* at the timestamp t^* satisfy the demand. The SM transforms his usage data

m^* timestamp t^* into a hash value μ: $\mu = \mathcal{H}_1(m^*\|P\|P_{pub}\|G\|H\|Q\|\ t^*)$, and hides its secret information x into element T': $T' = x\mu$. The SM proves that it has the knowledge of x. The SM randomly picks $k_5 \leftarrow \mathbb{Z}_p^*$, and computes $A = (k_5P + Q, P)$, $B = k_5\mu$, $h = \mathcal{H}_2(m^*\|T'\|A\|B\|C\|P\|P_{pub}\|G\|H\|Q\|t^*)$, $s_5 = k_5 - hx$. The SM sends the proof (m, h, s_5, ID) to the EU.

Upon receiving the proof, the EU confirms that the signature σ^* and the curtailment proof $(m, T', h, s_5, \text{ID})$ have the same secret information x. The EU computes $A' = e(S, \alpha P + P_{pub})e(s_5P + (h' - 1)C, P)$, $B' = s_5\mu + h'T'$ and $\alpha = \mathcal{H}_2(\text{ID})$. The EU checks whether $h' = \mathcal{H}_2(m^*\|T'\|A'\|B'\|C\|P\|P_{pub}\|G\|H\|Q\|t^*)$ and $T = T'$ hold. If they hold and $(m - m^*)/m > \lambda$, the EU determines that the consumer curtailed his energy consumption, then sends incentive payments to the consumer for his cooperation.

5   Security Analysis

In this section, we state the security analysis. The analysis is divided into four classes: data integrity, authentication, reward support and strong anonymity.

5.1   Data Integrity

The integrity includes the integrity of anonymous reports and integrity of instructions in our IAA scheme. When the SM sends the energy consumption data to the EU, it cloaks its credential (x, α, S), and proves the credential with zero knowledge proof. During the demand response part, the EU's consumption instruction is signed by a Schnorr short signature [15]. Since the Schnorr short signature is provably secure under the Gap-DL problem in the random oracle model, the integrity can be ensured. As the result, IAA can make sure the integrity of the anonymous report and the instruction.

5.2   Authentication

The authentication of the IAA scheme is based on the $q-$SDH assumption. During the joining protocol, each SM's credential C is signed by the EU with the BBS+ signature. BBS+ signature has been proved against chosen plaintext attack under the $q-$SDH assumption in the standard model. According to the analysis of the integrity, the anonymous report protocol is secure. Therefore, the third party cannot produce a valid signature without the help of the EU.

5.3   Reward-Support

The voluntary response protocol is a identity-based signature scheme that derives from zero knowledge proof. It implies that the adversary cannot tamper the public key α, which is from the collusion resistant function $\mathcal{H}_{\text{ID}}$. According to the analysis of integrity, the adversary cannot produce a valid but illegitimate usage report to frame a legitimate consumer under the security of the zero knowledge proof. Therefore, from above aspects, IAA can make sure that the voluntary consumer can get corresponding rewards.

5.4 Strong Anonymity

The consumer's anonymity is based on the DDH assumption in our IAA scheme. A SM generates its energy usage data using its secure key x. The essence of the anonymous report protocol is to shuffle the credential (x, α, S) to a temporary one $(T, R, U, g, s_0, s_1, s_2, s_3)$. After that, the SMs send their messages and signatures through an anonymous network. Because $\mathcal{H}_1$ is a collision resistant function, under the DDH assumption, it is infeasible to decide whether two elements T and T_0 are generated using the same secret information x. As such, no one can trace the legitimate signature from an honest SM unless he knows the secret key. Hence, our IAA scheme satisfies the anonymity requirement.

6 Performance Analysis

In this section, we evaluate the computational cost and the communication overhead required by our IAA scheme, and compare it with some previous works.

Table 1. Computational performance

	Party	Computational cost	Mean	Deviation	95% confidence interval
Set up	EU	$4G_p + G_m$	37.48ms	0.69ms	[37.15ms, 37.81ms]
Joining	EU	G_m	3.76ms	0.32ms	[3.60ms, 3.91ms]
	SM	$2G_p + 2G_m$	18.74ms	0.35ms	[18.57ms, 18.90ms]
Report	SM	$G_p + 3G_e + 8G_m$	31,00ms	0.68ms	[30.68ms, 31.32ms]
	EU	$2G_p + 4G_e + 8G_m$	48.62ms	0.74ms	[48.27ms, 48.98ms]
Demand	EU	G_m	3.35ms	0.13ms	[3.29ms, 3.41ms]
	SM	$2G_m$	4.57ms	0.10ms	[4.52ms, 4.62ms]
Response	SM	$G_p + 2G_m$	12.58ms	0.18ms	[12.50ms, 12.67ms]
	EU	$2G_p + 5G_m$	34.91ms	0.75ms	[34.55ms, 35.26ms]

6.1 Computational Cost

Firstly, we discuss the computational cost in our IAA scheme. Compared with exponentiation G_e, multiplication G_m and pairing evaluations G_p, the overheads of hash evaluations and arithmetic operations are very small. We emulate the scheme IAA on a Ubuntu 12.04 virtual operation system with a Intel Core i5-4300 dual-core 2.60 GHz CPU. We only use one core and 1 GB of RAM. To achieve 80 bits security level, we set the length of $\mathbb{G}$ to 161 bits and p to 160 bits. Some bilinear pairing operation can be calculated in advance. The computational costs and simulation results are presented in Table 1.

Secondly, we compare the Anonymous report's computational cost variation in terms of the number of SMs with conditionally anonymous ring signature (CRS) [19] and deniable ring signature (DRS) [20], which can also achieve similar

security properties. The comparison is based on PBC cryptography libraries [21]
and MIRACL libraries [22]. Fig. 1 shows the comparison result of compuational
cost for an anonymous report between a SM and the EU. According to the
figure, it can be seen that the computational cost is constant in IAA. Instead,
the computational cost and the number of SMs are directly related in CRS and
DRS.

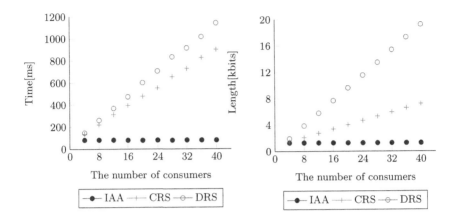

Fig. 1. Computational cost   **Fig. 2.** Communication overhead

6.2   Communication Overhead

In this subsection, we discuss the communication overhead between a SM and
the EU. In the joining procedure, a SM sends a credential and its identity to
the EU in form of $C\|ID$, whose length is $\|\mathbb{G}\| + \|ID\|$. During the anonymous
report procedure, the EU reports the consumption data m with the timestamp
t and signature σ, which is in form of $T\|R\|U\|g\|s_0\|s_1\|s_2\|s_3$. The size of σ is
$3\|\mathbb{G}\| + 5\|p\|$. In the Demand Generation protocol, the form of the signature
is $f\|s_4$, whose size is $2\|p\|$. After the SM curtails the energy consumption, it
sends the proof to the EU for asking the rewards. The form of the signature is
$T'\|s_5\|h\|ID$, whose size is $\|\mathbb{G}\| + 2\|p\| + \|ID\|$. Here, we compare the signature
size among IAA, CRS and DRS. The result is depicted in Fig. 2.

According to the Fig. 2, it can be seen that the communication complexity
is $O(1)$ in our scheme. Compared with CRS and DRS, whose communication
complexity is $O(q)$, where q is the number of SMs in the system, the number of
smart meters will not affect the communication cost in IAA.

In the both ring signature schemes, the authentication is based on the DL
assumption. A SM must utilize other peer SM's public keys to cloak its identity.
This requires that the SM calculates the signature for all SMs' public keys. The
authentication of IAA is based on the $q-$SDH assumption. The SM produces

its commitment. Then, the EU generates the credential to authorize the commitment. The anonymous report part employs the non-interactive zero knowledge proof to cloak the SM's credential. Therefore, the communication and computational cost is constant in IAA.

7   Conclusion

In this paper, we propose an incentive-based anonymous authentication scheme for demand-response management in Smart Grids. Our scheme guarantees the cooperative consumers can confirm their cooperation without harming the privacy of other consumers. The security analysis has demonstrated that our IAA scheme can achieve data integrity, identity authentication, reward support and anonymity simultaneously. According to the performance analysis, it can be seen our scheme has more advantage over the existing conditionally anonymous authentication schemes in terms of computation and communication overhead for Smart Grid systems. Therefore, we conclude our scheme solves the challenge of trading off between performance and security. However, the cooperative consumers have to revoke their anonymity to prove their curtailments. This leaks cooperative consumer's privacy at the power shortage time. For the future work, we will explore the new technologies to improve the IAA and provide the anonymous cooperation proof to the consumers.

References

1. Litos Strategic Communication, The Smart Grid: An Introduction, Tech. rep., U.S. Department of Energy, pp. 7, 14–19, 22 (2008)
2. Yarali, A., Rahman, S.: Smart grid networks: Promises and challenges. Journal of Communications 7, 409–417 (2012)
3. Eckert, C., Krau, C., Schoo, P.: Sicherheit im Smart Grid - Eckpunkte fuer ein Energieinformationsnetz. Stiftung-Verbundkolleg/Projekt Newise (90) (2011)
4. Efthymiou, C., Kalogridis, G.: Smart grid privacy via anonymization of smart metering data. In: First IEEE International Conference on Smart Grid Communications (SmartGrid-Comm), pp. 238–243. IEEE (2010)
5. Fan, Z., Kalogridis, G., Efthymiou, C., Sooriyabandara, M., Serizawa, M., McGeehan, J.: The new frontier of communications research: smart grid and smart metering. In: Proceedings of the 1st International Conference on Energy-Efficient Computing and Networking, pp. 115–118. ACM (2010)
6. McDaniel, P., McLaughlin, S.: Security and privacy challenges in the smart grid. IEEE Security & Privacy 7(3), 75–77 (2009)
7. Lo, Y.-L., Huang, S.-C., Lu, C.-N.: Non-technical loss detection using smart distribution network measurement data. In: Innovative Smart Grid Technologies - Asia (ISGT Asia), pp. 1–5. IEEE (2012)
8. Shamir, A.: Identity-based cryptosystems and signature schemes. In: Blakely, G.R., Chaum, D. (eds.) CRYPTO 1984. LNCS, vol. 196, pp. 47–53. Springer, Heidelberg (1985)
9. So, H.K.-H., Kwok, S.H.M., Lam, E.Y., Lui, K.-S.: Zero-configuration identity-based signcryption scheme for smart grid. In: IEEE First IEEE International Conference on Smart Grid Communications (SmartGridComm), pp. 321–326 (2010)

10. Nicanfar, H., Jokar, P., Beznosov, K., Leung, V.C.: Efficient authentication and key management mechanisms for smart grid communications. IEEE Systems Journal (6), 1–12 (2013)
11. Li, H., Liang, X., Lu, R., Lin, X., Shen, X.: EDR: An efficient demand response scheme for achieving forward secrecy in smart grid. In: IEEE Global Communications Conference (GLOBECOM), pp. 929–934 (2012)
12. He, D., Chen, C., Bu, J., Chan, S., Zhang, Y., Guizani, M.: Secure service provision in smart grid communications. IEEE Communications Magazine 50(8), 53–61 (2012)
13. Chu, C.-K., Liu, J.K., Wong, J.W., Zhao, Y., Zhou, J.: Privacy-preserving smart metering with regional statistics and personal enquiry services. In: Proceedings of the 8th ACM SIGSAC Symposium on Information, Computer and Communications Security, pp. 369–380 (2013)
14. Goldwasser, S., Micali, S., Rackoff, C.: The knowledge complexity of interactive proof systems. SIAM Journal on Computing 18(1), 186–208 (1989)
15. Schnorr, C., Claus, P.: Efficient signature generation by smart cards. Journal of Cryptology 4(3), 161–174 (1991)
16. Au, M.H., Susilo, W., Mu, Y.: Constant-size dynamic k-TAA. In: De Prisco, R., Yung, M. (eds.) SCN 2006. LNCS, vol. 4116, pp. 111–125. Springer, Heidelberg (2006)
17. Nguyen, L., Safavi-Naini, R.: Efficient and provably secure trapdoor-free group signature schemes from bilinear pairings. In: Lee, P.J. (ed.) ASIACRYPT 2004. LNCS, vol. 3329, pp. 372–386. Springer, Heidelberg (2004)
18. Chen, L., Ng, S.L., Wang, G.: Threshold anonymous announcement in VANETs. IEEE Journal on Selected Areas in Communications 29(3), 605–615 (2011)
19. Zeng, S., Jiang, S., Qin, Z.: An efficient conditionally anonymous ring signature in the random oracle model. Theoretical Computer Science 461, 106–114 (2012)
20. Komano, Y., Ohta, K., Shimbo, A., Kawamura, S.-i.: Toward the fair anonymous signatures: Deniable ring signatures. In: Pointcheval, D. (ed.) CT-RSA 2006. LNCS, vol. 3860, pp. 174–191. Springer, Heidelberg (2006)
21. Lynn, B.: PBC library, http://crypto.stanford.edu/pbc/
22. Multiprecision integer and rational arithmetic c/c++ library, http://www.shamus.ie/

Mediated Community from an Intergroup Perspective: A Literature Review

Vilma Lehtinen[1]([✉]), Eeva Raita[1,3], Mikael Wahlström[2], Peter Peltonen[3], and Airi Lampinen[1,4]

[1] Helsinki Institute for Information Technology HIIT, Espoo, Finland
`vilma.lehtinen@helsinki.fi`
[2] VTT Technical Research Centre of Finland, Helsinki, Finland
[3] Department of Social Research, University of Helsinki, Helsinki, Finland
[4] Mobile Life Centre, Stockholm University, Stockholm, Sweden

Abstract. The ways people organize themselves as communities shift along with the digitalization of social interaction. We review studies on mediated community to analyze which aspects of social interaction are considered to characterize community today. We elaborate on their scientific positioning, or as termed by Doise [14], levels of explanation from the intra-individual to the societal level. Noticing that viewing mediated community as an intergroup phenomenon has been marginal, we propose a research agenda that addresses mediated community explicitly from an intergroup perspective. To extend knowledge of how communities are formed and maintained in digitalized, networked settings, we encourage future research to better integrate this perspective, by focusing on (1) the ways in which outgroups contribute to a sense of community (2) the interaction that occurs on the borders of communities, and (3) the ways in which intergroup relations delineate the symbolic construction of communities.

Keywords: Community · Online · Virtual · Review · Intergroup

1 Introduction

Scholarly approaches to mediated community touch upon a core issue in social sciences: the characteristics of social interaction in changing societal settings. The goal of this paper is to review and organize academic understandings of what aspects of social interaction make technologically mediated communities into being. "Online" and "virtual" are used commonly as prefixes to the term community with the purpose of denoting the use of digital and networked communication technologies. We use the concept of "mediated" as an umbrella term to cover these two prefixes.

This paper is focused on how mediated community is understood and investigated as an intergroup phenomenon. The intergroup perspective to mediated community is especially relevant now that scholars, activists and policy-makers alike are increasingly concerned with the dynamics of hostile behaviour online

© Springer International Publishing Switzerland 2015
T. Tiropanis et al. (Eds.): INSCI 2015, LNCS 9089, pp. 145–159, 2015.
DOI: 10.1007/978-3-319-18609-2_12

(e.g. [57]). Online hate groups, whose existence is based on the expression of hostility toward other groups, such as ethnic or sexual minorities [15], are extreme examples of ingroup formation based on intergroup comparisons. Others, such as networks of knitting blogs, may not seem to build their social identity on derogating outgroups, although they, too, can develop strong ingroup norms, as Wei [52] has demonstrated. Both the polarization of opinions [57] and the potential to use online interaction in improving intergroup relations [20] have been identified in prior work. Postmes and Baym [42] note these possibilities, arguing that the Internet changes the characteristics of intergroup contact so critically that its role in affecting intergroup relations cannot be overlooked.

Communities in general can be understood as groups in a social psychological sense: they are collections of individuals who either interact with each other, are interrelated in a micro-social structure, or share a common experience or destiny [6]. Brown's [6] definition of a group encapsulates the current understanding of the concept in social psychology in stating that "a group exists when two or more people define themselves as members of it and when its existence is recognized by at least one other". Thus, the members of a group are not the only ones who have a say in how the group is perceived. People external to a group can also contribute to the formation of the group, challenge its value, or even deny its existence. While being discriminated on the basis of one's group membership may lead to efforts to quit the group [5], an external threat to a group's identity may strengthen the groups cohesiveness by encouraging its members to join forces against a common enemy [46].

To elaborate on how mediated community is explained in current social scientific literature and how well the literature captures it as an intergroup phenomenon, we review approaches to mediated community in four academic journals in the field of social sciences and Internet research. To categorize these approaches, we use Doise's [14] framework of the levels of explanation in experimental social psychology. The framework includes four levels: 1) the intra-individual level 2) the inter-personal and situational level 3) the positional level and 4) the ideological level. With the intra-individual level, Doise refers to a focus on the phenomena that can be measured by analyzing the individual, such as the changes in an individual's emotions or opinions. The inter-personal level and situational level extends this view to phenomena occurring either between individuals or within a situation, such as interaction within a group. The positional level includes examinations of the social categories that participants in an interaction situation represent, such as gender or social status. Studies on the ideological level aim to identify shared values, norms and understandings, and to examine how these are formed in interaction. Doise emphasizes that these levels are analytical distinctions that researchers make to focus their studies, not straightforward reflections of reality.

None of these levels maps directly onto the intergroup perspective. However, the positional level points towards this direction: it frames interaction as an intergroup situation, as participants are considered to behave and be treated as members of certain groups. In the following, we review prior literature on mediated communities, assessing the relative prevalence of these four levels across studies. We examine the extent to which the intergroup perspective is already embedded in the reviewed research on different levels of analysis. Our analysis indicates that, despite of the aforementioned acknowledgements of the importance of an intergroup approach, the perspective has been largely overlooked in prior literature. In addition to organizing existing knowledge, we point to future directions for research by suggesting how the state of the art could be extended to better consider the integroup perspective. We argue that including and articulating an intergroup viewpoint when theorizing mediated communities will advance and complement current understandings of their dynamics.

2 Scope of the Review

To review the levels of explanation on which mediated community has been approached in social sciences and Internet research, we focused our initial inquiry on four databases that cover publications in the fields of "social sciences" and "communication studies": ISI Web of Science: Social Sciences Citation Index (SSCI); ProQuest: Sociological abstracts, and EBSCO: SocINDEX with Full Text, and Communication & Mass Media Complete. We considered these four databases suitable for our purposes of identifying key journals in which social scientific discussion on mediated community takes place. The aim of our selective review is to demonstrate the relative prevalence of different approaches in studies of mediated community, not to provide a comprehensive review of all research on mediated community. We conducted the queries in January 2015.

2.1 Selection of Journals

Aiming to focus our enquiry on those journals where research on mediated community appears most frequently, we, first, queried the selected databases for peer-reviewed journal articles that include the term "online community" in their abstract. The journals for which the search returned most hits were (query hits per journal in the databases are presented in parentheses): New Media and Society (34 hits); Computers in Human Behavior (27 hits); Journal of Computer-Mediated Communication (18 hits) and The Information Society (16 hits). We then searched the selected four journals for peer-reviewed articles that include the term "community", with or without prefixes, in their abstract. This was done to identify studies focused on community in mediated interaction, regardless of the exact language the authors adopt (that is, whether the prefix to community is "online", "virtual", or non-existent). These searches provided us with an initial set of 479 articles, including 123 articles from New Media and Society, 178 from Computers in Human Behavior, 90 from Journal of Computer-Mediated Communication, and 88 from The Information Society.

2.2 Exclusion Criteria

The search for the term "community" in the abstracts returned many articles that did not consider mediated community in particular. To keep the review focused on mediated community, we excluded altogether 215 of the 479 initially selected articles from further analysis. Each author first assessed a subset of the initial selection of articles, and pinpointed those that they deemed should be excluded as irrelevant to the topic. In discussing these suggestions, we identified four common themes that formed our final criteria for exclusion.

We excluded articles in which the term community referred solely to one (or several) of these four options: (1) topics other than community as an entity (such as community volunteers) (n=21), (2) community as a data source for studying other topics, or an aspect to be brought up only in the implications (n=54), (3) community solely as a geo-located entity (n=51), or (4) community solely as a demographic, occupational, or religious entity (n=89) in cases where the article did not study or otherwise elaborate on how these entities interact online. After agreeing on the exclusion criteria, each author evaluated a subset of the articles. As a result, we included 264 articles on mediated community, spanning the period from 1981 to January 2015. The set includes 73 articles from New Media & Society, 99 from Computers in Human Behavior, 62 from Journal of Computer Mediated Communication, and 30 from Information Society.

2.3   Analysis

To identify approaches to mediated community from the selected articles, regardless of what possible prefixes to community might be used, we inquired the articles with the search term "communit*". We reviewed the articles for both the theoretical conceptualizations of mediated community and the associated methods of empirical investigation to define the level(s) of analysis that each study represents. At this point, we excluded a further 104 articles that did not explicitly define mediated community. Also, another 36 articles were excluded since although they defined mediated community, they did not cover any empirical data. Thus, we were left with a final set of 124 articles that did define mediated community and that were based on empirical material.

The first author initially reviewed the articles independently, after which all authors discussed together the viability of the resulting categorizations. The analysis was an iterative process in that after discussion among all authors, the first author revisited the work, consulting the other authors in borderline cases. As Doise [14] pinpoints, a single study can reflect multiple levels of analysis. We kept this option open in our review as well but through the iterative analysis process we did in the end position each study on only one level explanation. We will discuss the principles used to achieve this positioning as we present the findings.

3   Findings

We now present how we applied Doise's framework to our review set, considering research on mediated community on each level of explanation in turn. We illustrate how the levels of explanation occur in the reviewed publications, and how the levels differ from each other. The qualitative examination of our review set enables us to elaborate on the different viewpoints that are manifest in studies of mediated community, as well as to consider how these approaches could be extended to better address the intergroup perspective. Finally, we summarize how the articles in our set are distributed across the four levels. While our review set was otherwise equally divided across different levels of explanation, very few studies were placed on the positional level.

3.1   Intra-individual Level

Starting from the intra-individual level, we first identified expressions that we interpreted to analyze mediated community from the viewpoint of an individual. We positioned on this level 35 studies that conceptualize and study mediated community through individuals' sentiments, traits, and perceptions. These publications focus both on how participation in mediated communities affects individuals' intrapersonal processes, and vice versa, the effects of intra-individual factors on the probability of participating in mediated communities. Most (n=33) assessed these aspects with quantitative self-report measures, except for three that applied a qualitative approach. The qualitative accounts address individual motivations to join a certain community, and identify aspects of mediated interaction that are related to these motivations, including reasons for joining [44] as well as personal experiences [12] and goals [27].

The majority of the self-report measures approach mediated community through the notion of a "sense of community" or related concepts. The sense of community measure was originally developed by McMillan and Chavis [37] for face-to-face communities. Seven publications used the original measure or its revised versions [39], [3], including questions such as "I think my community is a good place for me to live". Blanchard's [3] revised version applies especially to mediated settings: the concept of a "sense of virtual community" is defined as "members' feelings of membership, identity, belongingness, and attachment to a group that interacts primarily through electronic communication". We also identified related constructs that we considered to represent an intra-individual point of view to mediated community, including a sense of belonging [38], feelings of community [28], perceived group cohesion [21] and user affinity [58]. Each aims to identify mediated community by analyzing individuals' experiences in relation to their perceptions of the mediated interaction in which they participate. To conclude, studies we included on this level approach mediated community as the sentiment and perception of an individual. While acknowledging emotions as a component of mediated community, this view neglects the role outgroups can play in shaping these emotions.

3.2   Inter-personal and Situational Level

We positioned publications that aim to capture inter-personal phenomena instead of individuals' sentiments to the inter-personal and situational level, even though self-reports were at times used as a method of investigation in these publications, too. Moreover, we included on this level a set of experimental studies in which features of interaction were manipulated, and observations of interaction occurred online. We positioned a narrow majority of the reviewed publications (n=45) to this level of analysis. A strong undercurrent that we identified in this level was the idea of mediated community as networks. A prominent advocate of this approach is Canadian sociologist Barry Wellman [53] who argues that networks would be an illustrative term to describe social bonds in contemporary societies, suggesting it as an improvement over place-bound, detached, affectionally-laden notions of community. In brief, mediated communities are viewed here as a set of connections between individuals. These connections can be considered as social capital, that is, a repository of relationships that can be drawn upon for different forms of support when needed.

The publications representing the network view employ either survey measures assessing the amount, type or strength of interaction ties, or observations of actual use by analyzing logged use data. The studies using surveys (n=15) use measures of social capital and related constructs to identify mediated community. For example, Ellison et al. [17] adapt Williams' [54] scales to assess Internet-specific social capital. The authors then compare these measures to the reported use of a social network site. This way of identifying mediated community as individuals' perceptions partly overlaps with the intra-individual approach. However, we distinguish these two levels based on how much they emphasize the interaction between individuals and individual sentiments directed towards a more abstract entity of mediated community, respectively.

The emphasis on interaction between individuals is more apparent when mediated communities are identified by observing the interaction itself. A prevalent method (n=14) for studying the structure of interaction was analyzing logged data. This could take the form of either examining message content or conducting social network analysis where researchers leverage hyperlinks connecting participants and the networks of messages that they send to each other, focusing on the frequency and direction of contact rather than on their content.

Message content, such as keywords [9], clusters of words that appear in user profiles [8], [43] and interrelatedness of message topics [40] is also used to identify mediated community. A further group of 12 studies studied the content of online interaction in a more qualitative manner. Rather than quantifying the interaction taking place online, these studies applied qualitative methods such as conversation analysis [19], case studies [55] and ethnography [51] to identify connections between individuals that could be interpreted to constitute mediated communities. Moreover, we positioned on this level three experimental studies in which characteristics of interaction (moderation [56], persuasive strategies [22] and social control [36]) were manipulated.

Finally, the inter-personal and situational level includes studies approaching mediated communities as "communities of practice" that are seen to exist whenever individuals collaborate around knowledge, regardless of how its members feel or talk about the community. Seven studies in our review set explicitly referred to this concept that was originally developed by Lave and Wenger [32]. In these cases, mediated community was studied with both quantitative and qualitative methods, and it was expected to be identified by examining the ways goals and knowledge are shared. Four further studies employ concepts that we consider to resemble this practice approach. In these, individuals collaborate around shared (unintended) interests (e.g. [26]) or around consumption (e.g. [30]). Although the approach pinpoints the importance of collaborative action in the formation and maintenance of a community, it neglects the role that people external to the community play in these efforts. Thus, it fails to leverage the intergroup perspective.

Moreover, while smaller groups can be identified within the networks, group identification is not a concern for the network approach, either. Studies representing the network approach pinpoint the connections that form communities in online interaction. Simultaneously, they do not extend the perspective to the gaps that keep some networks separate from each other, nor to the importance of these gaps for keeping a single network cohesive. While this approach has the potential to be extended beyond the boundaries of a single community, accomplishing such broadening would require a focus on the boundaries of the network by examining the limits beyond which resources do not flow. An exceptional example of this, drawn from our review set, is Kazmer's [29] study on the activities related to disengaging from an online learning environment. With a focus on disengagement from these spheres of "people who share activities, space, and technology, and who communicate with one another", Kazmer [29] points towards an intergroup perspective by illustrating the tensions related to a change in an individual's social position.

3.3 Positional Level

Studies we placed on the positional level of explanation view mediated communities not solely from within the group, but also as situations in which individuals representing different social positions interact with each other. The positional level of explanation has by definition the greatest potential to implement an intergroup perspective to mediated community, as studies representing this level focus on social positions, such as roles, status, or membership in a particular group. These social positions are acknowledged to have an impact on how people orient towards each other. However, as we moved from considering individuals and the interaction between them to addressing memberships of social categories, the amount of fitting publications in our review set dropped sharply.

We identified only five publications to represent this approach from within our review set. Three of these publications use experimental methods to investigate how the salience of an individual's group identity affects self-presentation [45], recommendation intentions [33] or group-orientation in the content posted [10]. In these studies, the salience of the participants' group identity was primed by changing the content on Web sites or by having the participants to read a short story. Jahnke's [25] field study was the only qualitative study we included on this level. It examines the construction of social categories in mediated interaction by focusing on the diverse roles and social structures that emerge in a mediated community over the years. Bjarnason [2], on the other hand, used a survey to study how social stratification in adolescent communities is reflected in mediated communities.

The clear minority of the studies in our review can be considered to represent the positional level of analysis, and even the articles included on this level do not put the intergroup perspective forward very explicitly. For instance, none of the studies manipulate group identities or social positions in order to cause conflicts between social groupings. Judging by the difficulty of identifying studies that fit the description of the positional level, approaching mediated community as a result of different social positions seems to be an underrepresented perspective in current literature.

3.4   Ideological Level

The studies included on this level (n=39) do not presume that mediated community would exist as an immutable entity towards which individuals would target their sentiments. Instead, they argue that individuals construct community together with each other, considering community as an outcome of negotiations over what kinds of shared values, norms, practices and the like are constitutive of the community. The most popular (used in 27 studies in our review set) method on this level was a qualitative analysis of message content. For example, this could take the form of discourse analysis or thematic categorization. Ethnographic approaches, such as participant observation, were another popular method (used in 9 studies): here, message content is just one part of what is observed in order to understand the processes of constructing community. Additionally, researchers might take part in the interaction in the community, observing activities and conducting interviews with participants.

Almost half of the publications on this level (n=18) were positioned as examples of the ideological level of explanation because they either analyzed the understandings that individuals participating in a mediated community share, or examined how these understandings are negotiated in interaction. For example, Matei [35] analyzed postings to the platforms associated with an early mediated community to address how the concept of a virtual community was negotiated within them. Interviews and ethnographic approaches, too, were used to assess shared, and at times conflicting, understandings. For example, Fernback [18] used interviews to pinpoint what the metaphor of community means in mediated contexts. In these eighteen publications mediated community is seen to be negotiated largely from within, and it is not explicit to which degree its boundaries are impinged upon from the outside.

Another half (n=21) of the studies included on this level, however, consider community not as a coherent entity but as different groups which struggle for their existence "inside" it. This approach can be considered to implement the intergroup viewpoint as it focuses on how groups are defined in relation to other groups. These groups may become defined by a variety of aspects, including gender [4], appropriate or inappropriate behaviour [49], legitimate aesthetics [11] or activism [48]. The focus is thus on how the boundaries between these groups are created through negotiations over what kinds of behavior is accepted from its members, and how different kinds of activities are valued. Straightforward intergroup conflict is identified in Phillips' [41] early study that covered the practices that long-time members of a newsgroup used to defend the group's boundaries from newcomers. As the system did not technically support banning, the long-time members communicated using inside jokes, educated newcomers, or appealed to outsiders, such as service providers, to preserve their definition of the community.

Finally, some of the studies we considered to represent this approach investigate how boundaries can be crossed. Lingel and Naaman [34] show how uploaded videos of pop and rock concerts connect not just individual users and fan communities, but also fan communities to bands. Akoumianikis [1] exemplifies that certain community-specific ways of using language and visual symbols may function, not as a factor that separates communities from each other, but as a potential means of connecting them. In Akoumianikis' [1] study, a shared online platform for tourism professionals, including features such as visualizations of activities in a vacation package, functioned as a way to develop a shared understanding between the professionals and to encourage collaboration. Explicating the ways in which ingroup boundaries are defended, and at times, crossed, these studies acknowledge outgroups as contributors in the process. Mediated community is in these cases understood as a process of constant negotiation that occurs on its permeable boundaries.

3.5 Discussion

The distribution of articles in our review set on the different levels of explanation is quite equal, with the exception of the underrepresented positional level. Only five publications were placed on it, while we identified 35 instances of the intra-individual level, 45 of the situational level, and 39 of the ideological level. Applying Doise's framework of the levels of explanation in experimental social psychology for this review allowed us to discern the relative prevalence of different approaches to mediated community and to determine gaps in this domain. The most apparent finding is the relative scarcity of research on the positional level of analysis. While such studies may exist outside our review set, our review of publications in four prestigious journals serves to indicate the relative distribution of currently dominant perspectives to mediated community.

4   The Potential for an Intergroup Perspective

Above, we observed that currently, mediated community is mostly conceptualized and investigated from the inside, that is, with a focus on the community itself. The contributions of individuals who are left outside of mediated communities, or on their borders, receive scarce attention. Addressing this shortcoming, our analysis provides a basis for extending studies on mediated community by taking the intergroup perspective into account. We see untapped potential on each of the four levels of explanation to expand knowledge of how mediated community evolves as an intergroup process. We will now focus on three themes that provide particularly promising starting points for examining mediated communities from an intergroup perspective: (1) outgroups contributing to intra-individual factors, such as the sense of community, (2) interpersonal interaction occurring on group boundaries and the flow of resources in social networks, and (3) intergroup relations delineating the symbolic construction of communities on the ideological level. We consider these three themes to represent Doise's levels of explanation, except the positional level that is, by definition, inherent in these approaches.

4.1   Outgroups Contribute to How the Group is Perceived by Its Individual Members

We found that the intra-individual level of explanation is emphasized, for example, in the prevalent notion of mediated community as a sense of community. Studies on how perceptions about outgroups contribute to a sense of community as experienced by individual members could further the knowledge of how emotions regarding mediated communities evolve. Focusing on the situation between an ingroup and an outgroup (rather than a view from within a group) could illustrate how intergroup relations may affect the dynamics of creating sentiments towards the ingroup. For example, as the theory of social identity [47] suggests, to build up a positive social identity, ingroup is favored even in the absence of any intergroup conflict over scarce resources. On the other hand, observed differences between groups can be used to justify acts of favoring the ingroup [50]. Lampinen et al. [31] provide an example of outgroups contributing to the ways the ingroup is perceived by its individual members. Their study illustrates how IT students, a group of people who interact with each other actively both online and offline, build up their sense of community by distinguishing themselves from the mainstream of their generation. This is done both discursively and with technology choices, for example, by using older, less accessible, text-based forms of online communication, such as the Internet Relay Chat (IRC).

Focusing on such intergroup processes could provide more detail on how mediated communities emerge in interactions between groups. The intergroup perspective would be crucial also to examine whether, how, and to what extent mediated communities differ from face-to-face communities in this respect. Assessing the formation of mediated communities as an intergroup phenomenon would reveal whether mediated communities are based more on emphasizing ingroup similarities than intergroup differences. These studies could also assess the consequences of mediated interaction for intergroup relations, such as the extent to which ingroup sentiments are based on outgroup derogation.

4.2   Consequential Interaction Occurs on Group Boundaries

The notions of mediated community as a network, and other approaches positioned on the interpersonal level, are mostly concerned on how resources flow within the network. The limits beyond which the resources do not flow typically remain beyond the scope of this approach. To take advantage of the intergroup perspective, the focus could be tilted towards the boundaries of resource sharing. Even if these boundaries are permeable and transient, they can be located somewhere at particular points in time. For example, studies on social stratification in late modern societies have shown that social capital (or the lack of it) flows across generations (see, e.g. [24]). Examining the gaps in the networks of resource sharing could explicate whether online networks are similarly affected by perceptions of who belongs to the network and who does not. It would also inform us about the formation of densely knit communities within a dispersed network, by illustrating how communities are juxtaposed with the broader network. Empirical studies of mediated communities of practice could, for example, focus their view on the practices that maintain the community, such as negotiations on whether one is expert or reputed enough to join a community.

By examining the boundaries that delineate communities, we could find out whether new boundaries are created online, or whether they are diminished altogether, and how. For example, studies on the so-called sharing economy [23], [16] have pinpointed an ambivalent orientation to members of outgroups on peer-to-peer exchange communities online. While being able to connect and exchange with new people is central to these services, people tend to be selective in choosing the "strangers" with whom they interact. At times, this takes the form of outright discrimination. An instance of the diminishing of boundaries, on the other hand, is the process of becoming an insider member of a blogging community through the actions of long-time members, such as when the long-time members link to newcomers' blogs from their own blogs [13].

4.3   Intergroup Relations Delineate the Symbolic Construction of Communities

We suggest that research on the ideological level, covering the interactional processes of constructing shared identities, should acknowledge participants' memberships in differing social groups, as well as negotiations over who is allowed to belong to a certain group. In our review set, Braithwaite [4] provides an example of how intergroup relations delineate the symbolic construction of communities. The study illustrates how the boundaries of a gaming community are policed from feminism with explicit anger at feminists and anxiety over masculinity on online forums associated with the game. The acknowledgment of intergroup relations is essential to unravel their effect on the dynamics of ingroup formation. As Brown [6] points out, groups always exist in relation to other groups and individuals. The ingroup is therefore not in full control of how it is constructed in symbolic interaction. Outgroups make their own constructions of it. These may affect how members come to view their group, especially in the case of less powerful minorities. Some of the studies we positioned on the ideological level already acknowledge how communities are constructed in relation to other communities, but the viewpoint remains still largely one from within the community. This means, for example, focusing on how the community guards its boundaries from outsiders. The power that outgroups have to define these boundaries are not addressed, although such efforts can range all the way to aggressive attacks.

We presume that a focus on outgroups contributions to the symbolic construction of communities would help detailing the dynamics of mediated communities.

5   Conclusions

From our literature review, we conclude that there is a shortage of studies of mediated community that adopt an intergroup perspective. Yet, we argue that there is strong potential to extend the state of the art by considering this perspective as a part of future analyses. An intergroup perspective could help in providing new understandings of how the boundaries of mediated communities evolve, and how communities develop to extend their presumed boundaries. Each of the levels of explanation provides insight to the dynamics of mediated community. However, the intergroup perspective should be articulated explicitly to better support the design of systems and policies related to mediated communities. In essence, the intergroup perspective helps to identify barriers to community-building which lie outside the community. On the intra-individual level, sentiments towards outgroups could be identified as a variable affecting how the ingroup is perceived. On the interpersonal and situational level, barriers to resource sharing could be located at the borders beyond which the sharing of resources does not extend. The potential of the positional level could be fulfilled if intergroup relations were acknowledged as a trigger for actions occurring within a community. On the ideological level, the intergroup perspective shows how intergroup relations delineate the discursive boundaries of a community.

Mediated communities differ in the degree to which their formation is based on juxtapositions between communities. In online hate groups [15], hostility towards other groups is far more explicit than in knitting blogs [52]. However, regardless of the extent to which the ingroup identity of a mediated community is based on outgroup derogation, processes of ingroup formation always rely on intergroup comparisons. Social categories, ranging from music genres and household types, to ethnicities and social class, become meaningful only in relation to each other. Trash metal fans are distinguished from glam rockers, cohabiting couples from single mothers, and so on. Intergroup relations do not always culminate in conflict, but the potential for it lies in the human tendency to strengthen ingroup identities through intergroup comparisons [7]. Whether the boundaries between groups are considered as symbolically constructed, as feelings, or as observable interaction, people orient to their surroundings based on their group identities. We argue that taking this tendency into account is productive for the study of mediated communities. While mediated interaction can extend the notion of community beyond geographical locality, it may reinforce other division lines. Understanding mediated community as an intergroup process allows for examining how these divisions emerge as a result of intergroup relations.

References

1. Akoumianakis, D.: Ambient affiliates in virtual cross-organizational tourism alliances: A case study of collaborative new product development. Comput. Hum. Behav. 30, 773–786 (2014)
2. Bjarnason, T., Gudmunsson, B., Olafsson, K.: Towards a digital adolescent society? The social structure of the Icelandic adolescent blogosphere. New Media Soc. 13, 645–662 (2011)
3. Blanchard, A.: Developing a Sense of Virtual Community Measure. CyberPsychol. Behav. 10(6), 827–830 (2007)
4. Braithwaite, A.: Seriously, get out: Feminists on the forums and the War (craft) on women. New Media Soc. 16(5), 703–718 (2014)
5. Branscombe, N.R., Schmitt, M.T., Harvey, R.D.: Perceiving pervasive discrimination among African Americans: Implications for group identification and well-being. J. Pers. Soc. Psych. 77, 135–149 (1999)
6. Brown, R.: Group processes: dynamics within and between groups, 2nd edn. Blackwell, Oxford (2000)
7. Brown, R.: Prejudice, 2nd edn. Wiley-Blackwell, Chichester (2010)
8. Cantador, I., Castells, P.: Extracting multilayered Communities of Interest from semantic user profiles: Application to group modeling and hybrid recommendations. Comput. Hum. Behav. 27(4), 1321–1336 (2011)
9. Choi, S., Park, H.: An exploratory approach to a Twitter-based community centered on a political goal in South Korea. Who organized it, what they shared, and how they acted. New Media Soc. 16(1), 129–148 (2014)
10. Cress, U., Schwmmlein, E., Wodzicki, K., Kimmerle, J.: Searching for the perfect fit: The interaction of community type and profile design in online communities. Comput. Hum. Behav. 38, 313–321 (2014)
11. Cristofari, C., Guitton, M.: Mapping virtual communities by their visual productions: The example of the Second Life Steampunk community. Comput. Hum. Behav. 41, 374–383 (2014)
12. Crowson, M., Goulding, A.: Virtually homosexual: Technoromanticism, demarginalisation and identity formation among homosexual males. Comput. Hum. Behav. 29(5), 31–39 (2013)
13. Dennen, V.: Becoming a blogger: Trajectories, norms, and activities in a community of practice. Comput. Hum. Behav. 36, 350–358 (2014)
14. Doise, W.: Levels of explanation in social psychology. Cambridge University Press, Cambridge (1986)
15. Douglas, K.: Psychology, discrimination and hate groups online. In: Joinson, A., McKenna, K., Postmes, T., Reips, U. (eds.) The Oxford Handbook of Internet Psychology, pp. 115–164. Oxford University Press, Oxford (2007)
16. Edelman, B., Luca, M.: Digital Discrimination: The Case of Airbnb.com. Harvard Business School NOM Unit Working Paper No. 14-054 (2014)
17. Ellison, N., Steinfield, C., Lampe, C.: The benefits of Facebook friends: Social capital and college students use of online social network sites. J. Comput. Mediat. Comm. 12(4), 1143–1168 (2007)
18. Fernback, J.: Beyond the diluted community concept: a symbolic interactionist perspective on online social relations. New Media Soc. 9(1), 49–69 (2007)
19. Giles, D., Newbold, J.: Is this normal? The role of category predicates in constructing mental illness online. J. Comput. Mediat. Comm. 18(4), 476–490 (2013)

20. Hasler, B., Amichai-Hamburger, Y.: Online intergroup contact. In: Amichai-Hamburger, Y. (ed.) The Social Net: Understanding Our Online Behavior, 2nd edn., pp. 220–252. Oxford University Press, Oxford (2013)

21. Hsu, C., Lu, H.: Consumer behavior in online game communities: A motivational factor perspective. Comput. Hum. Behav. 23(3), 1642–1659 (2007)

22. Kim, H., Sundar, S.: Can online buddies and bandwagon cues enhance user participation in online health communities? Comput. Hum. Behav. 37, 319–333 (2014)

23. Ikkala, T., Lampinen, A.: Monetizing Network Hospitality: Hospitality and Sociability in the Context of Airbnb. In: Proc. CSCW 2015. ACM Press, New York (forthcoming, 2015)

24. Jaeger, M.: Educational mobility across three generations: the changing impact of parental social class, economic, cultural and social capital. Eur. Soc. 9, 4 (2007)

25. Jahnke, I.: Dynamics of social roles in a knowledge management community. Comput. Hum. Behav. 26, 233–546 (2010)

26. Josefsson, U.: Coping with Illness Online: The Case of Patients' Online Communities'. Inform. Soc. 21, 133–141 (2005)

27. Jung, Y., Kang, H.: User goals in social virtual worlds: A means-end chain approach. Comput. Hum. Behav. 26(2), 218–225 (2010)

28. Kahn, A., Ratan, R., Williams, D.: Why We Distort in Self-Report: Predictors of Self Report Errors in Video Game Play. J. Comput. Mediat. Commun. 19(4), 1010–1023 (2014)

29. Kazmer, M.: Beyond CU L8R: Disengaging from online social worlds. New Media Soc. 9(1), 111–138 (2007)

30. Lampel, J., Bhalla, A.: The role of status seeking in online communities: Giving the gift of experience. J. Comput.-Mediat. Comm. 12(2), 434–455 (2007)

31. Lampinen, A., Lehtinen, V., Cheshire, C.: Media Choice and Identity Work: A Case Study of Information Communication Technology Use in a Peer Community. Emerald Studies in Media and Communication. In: Robinson, L., Cotten, S.R., Schulz, J. (eds.) Communication and Information Technologies Annual 2014: Doing and Being Digital: Mediated Childhoods (2014)

32. Lave, J., Wenger, E.: Situated Learning: Legitimate Peripheral Participation. Cambridge University Press, Cambridge (1991)

33. Lee, D., Kim, H.S., Kim, J.K.: The role of self-construal in consumers electronic word of mouth (eWOM) in social networking sites: A social cognitive approach. Comput. Hum. Behav. 28(3), 1054–1062 (2012)

34. Lingel, J., Naaman, M.: You should have been there, man: Live music, DIY content and online communities. New Media Soc. 14(2), 332–349 (2012)

35. Matei, S.: From counterculture to cyberculture: Virtual community discourse and the dilemma of modernity. J. Comput. Mediat. Comm. 10, 3 (2005)

36. Matzat, U., Rooks, G.: Comput. Hum. Behav. Styles of moderation in online health and support communities: An experimental comparison of their acceptance and effectiveness. Comput. Hum. Behav. 36, 65–75 (2014)

37. McMillan, D., Chavis, D.: Sense of community: a definition and theory. J. Community Psychol. 14, 6–23 (1985)

38. Park, J., Gu, B., Leung, A., Konana, P.: An investigation of information sharing and seeking behaviors in online investment communities. Comput. Hum. Behav. 31, 1–12 (2014)

39. Peterson, N., Speer, P., McMillan, D.: Validation of a brief sense of community scale: confirmation of the principal theory of sense of community. J. Community Psychol. 36(1), 61–73 (2008)

40. Pfeil, U., Zaphiris, P., Wilson, S.: The role of message-sequences in the sustainability of an online support community for older people. J. Comput.-Mediat. Comm. 15, 336–363 (2010)

41. Phillips, D.: Defending the Boundaries: Identifying and Countering Threats in a Usenet Newsgroup. Inform Soc. 12(1), 39–62 (1996)

42. Postmes, T., Baym, N.: Intergroup dimensions of Internet. In: Harwood, J., Giles, H. (eds.) Intergroup Communication: Multiple Perspectives, pp. 213–238. Peter Lang., New York (2005)

43. Procaci, T., Siqueira, S., Braz, M., de Andrade, L.: How to find people who can help to answer a question? Analyses of metrics and machine learning in online communities. Comput. Hum. Behav. (in press)

44. Ridings, C.M., Gefen, D.: Virtual community attraction: Why people hang out online. J. Comput. Mediat. Comm. 10, 1 (2004)

45. Schwammlein, E., Wodzicki, K.: What to Tell About Me? Self Presentation in Online Communities. J. Comput. Mediat. Comm. 17(4), 387–407 (2012)

46. Scheepers, D., Spears, R., Doosje, B., Manstead, A.: The social functions of ingroup bias: Creating, confirming, or changing social reality. Eur. Rev. Soc. Psychol. 17, 359–396 (2006)

47. Tajfel, H., Turner, J.C.: The social identity theory of intergroup relations. In: Worchel, S., Austin, W.G. (eds.) Psychology of Intergroup Relations, pp. 7–24. Nelson-Hall, Chicago (1986)

48. Tatarchevskiy, T.: The popular culture of internet activism. New Media Soc. 13(2), 297–313 (2011)

49. Tiidenberg, K.: Boundaries and conflict in a NSFW community on tumblr: The meanings and uses of selfies. New Media Soc. (2015) (in press)

50. Turner, J., Hogg, M., Oakes, P., Reicher, S., Wetherell, M.: Rediscovering the social group: A self-categorization theory. Basil Blackwell, Oxford (1987)

51. Uotinen, J.: Involvement in (the information) society - the Joensuu community resource centre netcafe. New Media Soc. 5(3), 335–356 (2003)

52. Wei, C.: Formation of norms in a blog community. In: L. Gurak, S. Antonijevic, L. Johnson, C. Ratliff, J. Reyman (Eds.), Into the Blogosphere. Rhetoric, Community, and Culture of Weblogs (2004)

53. Wellman, B.: Physical Place and Cyberplace: The Rise of Personalized Networking. Int. J. Urban Regional 25(2), 252–252 (2001)

54. Williams, D.: On and off the net: Scales for social capital in an online era. J. Comput.-Mediat. Comm. 11(2) (2006)

55. Winston, E., Medlin, B., Romaniello, B.: An e-patients end-user community (EUCY): The value added of social network applications. Comput. Hum. Behav. 28(3), 951–957 (2012)

56. Wise, K., Hamman, B., Thorson, K.: Moderation, response rate, and message interactivity: Features of online communities and their effects on intent to participate. J. Comput. Mediat. Comm. 12(1), 24–41 (2006)

57. Yardi, S., Boyd, D.: Dynamic debates: an analysis of group polarization over time on Twitter. B. Sci. Technol. Soc. 30, 316–327 (2010)

58. Zha, X., Zhang, J., Yan, Y., Xiao, Z.: User perceptions of e-quality of and affinity with virtual communities: The effect of individual differences. Comput. Hum. Behav. 38, 185–195 (2014)

Author Index

Printed in the United States
By Bookmasters